Utku Ay

Large Format CMOS Image Sensors

Suat Utku Ay

Large Format CMOS Image Sensors

Performance and Design

VDM Verlag Dr. Müller

Imprint

Bibliographic information by the German National Library: The German National Library lists this publication at the German National Bibliography; detailed bibliographic information is available on the Internet at http://dnb.d-nb.de.

Cover image: www.purestockx.com

Publisher:
VDM Verlag Dr. Müller Aktiengesellschaft & Co. KG, Dudweiler Landstr. 125 a, 66123 Saarbrücken, Germany,
Phone +49 681 9100-698, Fax +49 681 9100-988,
Email: info@vdm-verlag.de

Produced in USA and UK by:
Lightning Source Inc., La Vergne, Tennessee, USA
Lightning Source UK Ltd., Milton Keynes, UK
BookSurge LLC, 5341 Dorchester Road, Suite 16, North Charleston, SC 29418, USA

ISBN: 978-3-8364-7052-0

to

my family

TABLE OF CONTENTS

LIST OF TABLES

CHAPTER 1 INTRODUCTION

Recently, many semiconductor companies around the world which produce complementary metal oxide semiconductor (CMOS) mixed-signal integrated circuits (IC) have been investing in the area of solid-state image sensors. On application side, these efforts have been driven preliminarily by consumer electronics applications such as cellular phones, digital still cameras (DSC), PC cameras, and camcorders that require image sensors. On technology side, CMOS image sensors have became more mature and attractive to be fabricated in mass quantities with relatively higher profit margins. On the business side, companies with strong position in the CMOS IC markets seize the opportunity to fill their access fabrication capacity with minimal investment with higher returns. Their ability to control their fabrication process and cost with their know-how in the area of mixed-signal IC design increased their chance to be successful and profitable in the area of solid-state image sensor markets meeting the market demand on larger array formats, high image quality, low power, low cost, small footprint imaging IC solutions.

Charge coupled device (CCD) technology has been the unequalled leader in the field of solid-state imaging for wide range of applications for over two decades. CCD technology had matured in terms of production yield and performance. In last decade, a relatively new image sensor technology called active pixel sensor (APS) [Fossum93a] that uses existing CMOS manufacturing facilities has emerged as a potential replacement of CCD. CMOS APS technology preserves almost all of the desirable features of CCD, yet circumvents major weaknesses of CCD technology. CMOS APS technology has several advantages over CCD technology, including; lower cost, lower power consumption (100 to 1000 times lower) [Cho00], higher dynamic range, higher blooming threshold, individual pixel readout, low supply voltage (2.8V, or lower) operation, high speed, large array size, radiation hardness, and smartness by incorporating on-chip signal processing circuits [Fossum93b].

High-resolution imaging applications such as professional photography, astronomical imaging, x-ray imaging, TV broadcasting and machine vision require very large format image sensors. CCD image sensors have been fabricated with very large array formats to support these applications. 66 Million-pixel, front-illuminated CCD with 12 μm pixel pitch by Philips [Lesser97], 85 Mpixel wafer scale CCD with 8.75 μm pixel pitch by Lockheed-Martin (now Loral) [Wen99], and 111 Mpixel single chip CCD with 9.0 μm pixel pitch by STA Inc [Bredthauer07] were introduced in 1997, 1999 and 2007, respectively. However, large format CCDs are very expensive and difficult to produce with the low defect densities needed for high quality imaging. When

1

increased pixel full-well capacity (well-depth) and higher spectral response requirements are added, the necessary increase in pixel size (and hence, the sensor size) makes the production of such CCDs prohibitively expensive. Furthermore, power consumption and need for external support electronics make CCDs less attractive for earth-based or space-borne high-resolution applications. Recently, CMOS APS technology, on the other hand, have gained more popularity in these image sensor segments with the recent advancement in frame rates [Olsen97, Krymsky99], noise levels and array formats [Foveon00, Meynants03, Iwane07, Forza07]. This was achieved by utilizing better image sensor architectures and design techniques, and by advancement in the CMOS fabrication processes and pixel technologies.

Spectral response of a solid-state image sensor is directly related to the physical and spectral properties of the photosensitive element as well as the overlaying layers on top of the imaging pixel. For example, in ultraviolet (UV) (30-390nm) and soft X-ray spectrums (1-10nm), a frontside illuminated CCD is insensitive because of the high quantum losses in the overlaying gate and oxide layers. Backside illumination has been used for extending and enhancing spectral response of CCDs in UV spectrum [Muramatsu93]. However, backside thinning, passivation, UV enhancement, packaging, and long-term stability issues have prevented stand alone use of these devices in UV spectrum applications. Currently down converting phosphor-coated frontside (or backside) illuminated CCDs with low quantum efficiency (QE) and modulation transfer function (MTF) are used in UV spectrum applications. UV response improvement of CCD for space-borne telescopes has been of great interest area of research activity [Groom00]. Yet, UV spectrum remains relatively uncharted territory for large format CMOS APS technologies. This is mainly because UV improvement is directly related to the quality of manufacturing process and pixel technologies. Until recently, almost none of the image sensor companies have control over the fabrication process they used. In addition, lack of funding, financial prospect, and market size could be counted as other difficulties facing researchers in this segment.

1.1 Motivation and Goals of This Book

Die size of a CMOS integrated circuit (IC) is limited by the exposure field size of the photolithographic stepper used during manufacturing which is typically 20mm by 20mm. However, by using modern photolithographic stepper equipment and/or techniques in CMOS IC manufacturing, an integrated circuit die area may easily exceed this limited exposure field size. So-called stitching technology [Tower99] allows designer to physically merge several design structures on the same wafer in order to make a large image sensor ICs, [Ferrier97, Kreider95, Malinovich99, Forza07]. The stepper exposes the entire image sensor structure, one piece at a time, by precisely aligning each reticle step. The stitching process places several constraints on the image sensor design. These constraints as well as image sensor's electro-optical design requirements have to be addresses in CMOS image sensor design methodologies for high performance and for first silicon success. In this work, current state-of-the-art CMOS image sensor design methodologies were investigated for fables design companies, and a stitching design methodology was proposed for large format CMOS image sensor design.

Image sensor pixel design is constrained by pixel full-well capacity, sensor resolution, wafer/die size, quantum efficiency, sensitivity, and dark current. Pixel size is limited by the reticle (die or wafer size) and quality of the supporting optics. Thus, large pixel full-well capacity could only be achieved through the use of novel fabrication process and circuit design techniques. Increased pixel full-well capacity leads higher dynamic range if combined with low noise readout which are the two most important requirements for scientific image sensors. Quantum efficiency (QE) mostly depends on the fabrication process technology and pixel design techniques. In photodiode type CMOS APS pixels, especially in near-UV spectrum (200-400nm), QE could be improved by novel pixel design techniques. Achieving high dynamic range by achieving very large pixel full-well capacity (or photodiode capacitance), at the same time improving near-UV spectrum response of APS pixels are the two chalanges that were addressed and investigated in this work.

Pixel array on the focal plane is the central but not the only important part of a CMOS image sensor (CIS) IC. Pixels are co-exist with mixed-signal circuits on the die. Today's mixed-signal design methodologies have to be modified to accomplish not only the analog and digital circuit quality but also the quality of the image sensor array in CMOS image sensor (CIS) ICs. Thus, new mixed-signal CIS IC design methodologies are needed. These methodology issues were address in this work along with the electrical and optical modeling of CMOS pixel parameters. CMOS pixel models were used to estimate pixel performance metrics during the design and technology evaluation processes.

Detailed circuits and pixel design issues were dicussed in different chapters with case studies to give reader a chance to appreciate steps involved in designing and characterizing a CMOS image sensor.

1.2 Organization of Book

This book organized in eight chapters. Motivation and goals were summarized in the first chapter. In second chapter, common properties and methods involving in solid-state image sensing process were described. Main performance criteria and definitions of silicon bases image sensors, measurement and characterization methods were explained in chapter two. In third chapter, state-of-the-art CCD technologies and their general properties were reviewed. Techniques that have been used in CCD technologies for high quality image sensing and scientific applications were also investigated in this chapter. Proposed CMOS process and design rule based active and passive pixel models are explained in the chapter four. State-of-the-art CMOS image sensor technologies were reviewed in chapter four as well. A prototype CMOS image sensor platform design as a pixel evaluation and development platform was introduced in fifth chapter. Chapter six deals with the pixel full-well capacity and quantum efficiency improvement methods for photodiode and hybrid type CMOS APS image sensors. Two methods were investigated for increasing pixel full-well capacity, and for improvement of near-UV response of CMOS APS pixels. Design methodologies to map CMOS fabrication process, technology, and design parameters onto system and image sensor performance requirements to achieve optimum performance with CMOS image sensors were reviewed in chapter seven. In chapter seven, a methodology was proposed to build a very large CMOS APS sensor chip. A case study of building very large, single die, CMOS APS image sensor with 16.85 million-pixel and 1.35 million electrons full-well capacity was presented with detailed circuit, and layouts, test and characterization results in chapter eight along with conclusion and the future prospects of large format image sensors.

CHAPTER 2 SOLID STATE IMAGING - BACKGROUND

Photon collection and imaging process is manifested by certain physical principles. What makes it challenging is that the choices an image sensor designer has to make for harvesting impinging photons efficiently. Fundamental design issues, working principles, characteristics, and measurement methods of silicon based image sensors are discussed in this chapter.

2.1 Solid State Imaging Process

Solid-state imaging is based on a physical process of converting one energy type into another measurable quantity. Input energy is the quanta of light energy, or photon energy. Output measurable quantity is electric current or voltage. Link between these two is the electron that is generated in a medium where photon energy is used to push the electron from its valance band to conduction band in where it can travel freely leaving a virtual "hole" behind. Important part of the imaging process is to collect these free electrons (or holes) in confined areas, so called pixels, and transfer them to a location in where they can be processed and converted into meaningful electrical quantities. Principle steps of this process are depicted in Figure 2.1.

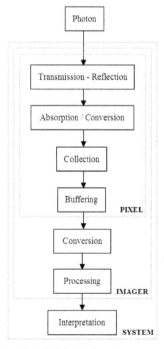

Figure 2.1. Solid-state imaging process.

5

Imaging process starts at pixel. Impinging photons pass through dielectric layers, absorbed in pixel structures and converted into charges, electrons or holes. These charges are then collected in a three dimension (3D) confined areas, and buffered or read sequentially to upper level of processing circuits. Image sensor converts pixel signals into more meaningful signal types, and processes them such a way that today's signal processors could use and transport them. Processing circuits convert and process these pixel readings to form one (1D), two (2D), or three (3D) dimensional scene images. In system level, these images are again processed or interpreted for human or machine use. All these processes are done at different level of hierarchies as shown in Figure 2.1. Although borders of hierarchies were drawn clearly in the figure, these boundaries are not defined as strictly.

2.1.1 Transmission and Reflection Process

Photons pass through multi layers of dielectrics in modern solid-state image sensors before reaching to photo conversion sites. These dielectric layers are placed on top of solid-state material to isolate different functional layers, such as multi layer routing metals. Some of these layers are opaque and some of them are transparent. Because each layer has different optical properties, some portions of the impinging photons are reflected, and some are absorbed leading to quantum loss. Velocity of a light in a medium is given by a wavelength-dependent complex refractive index, N_j.

$$N_j = n_j - i \cdot k_j \qquad [2.1]$$

$$c_j = \frac{c_0}{N_j} \qquad [2.2]$$

$$\frac{1}{c_j} = \frac{n_j}{c_0} - \frac{i \cdot k_j}{c_0} \qquad [2.3]$$

Real part of equation [2.1] (n_j) is related to the velocity of light in medium j. It governs normal incident dispersion. Term k_j is the extinction coefficient. It governs absorption of light in the medium j. c_j is the speed of light in medium j, and c_0 is the speed of light in vacuum. Extinction coefficient is related to the absorption depth of material j with equation [2.4].

$$k_j = \frac{\lambda_j}{2 \cdot \pi \cdot \ell} \qquad [2.4]$$

ℓ is the absorption length of light with wavelength of λ_j in the medium j.

Reflection and transmission of incident plane waves from dielectric layers can be easily derived by using Fresnel equations [Balanis89] and by applying boundary conditions on the dielectric surfaces using the Maxwell equations [Balanis89]. This is very straightforward and could

6

be easily implemented if number of layers is small. However, this task becomes very cumbersome when the number of dielectric layers is increased. It was shown [Groom99, Pedrotti93, Anemogiannis92] that simple rearrangement of the boundary condition equations applied to one dielectric slab surface results in a transfer matrix formulation for the reflection and transmission coefficients that could be extended to stacked dielectrics slabs. Critical parameters that were used to determine the transfer matrix formulation of a dielectric slab are shown in Figure 2.2.

Dielectric slab was characterized by a transfer matrix formulation, containing layer thickness, input angle, and the phase shift due to the refractive index differences of layers. The phase shift of transmitted or reflected light from one medium to another is given by following equation.

$$\Phi_j = \frac{2 \cdot \pi \cdot d_j}{\lambda_0} \cdot N_j \cdot \cos\theta_{tj} \tag{2.5}$$

where λ_0, d_j, θ_{tj} and N_j are free space wavelength of the impinging light, thickness of the dielectric slab, transmission angle, and complex refractive index of dielectric slab, respectively. Snell's law relates transmission angle (θ_{tj}) to incident angle (θ_{j-1}). It is given in the equation [2.6] [Balanis89].

$$n_{(j-1)} \cdot \sin\theta_{t(j-1)} = n_j \cdot \sin\theta_{tj} \tag{2.6}$$

2x2 transfer matrix (M_j), relating input and output field strengths, could be obtained by solving electromagnetic boundary value problem for a single dielectric slab. Result is given in equation [2.7].

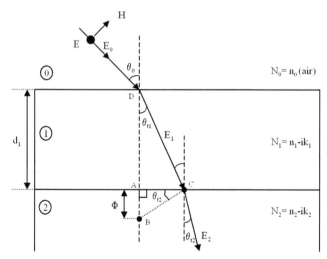

Figure 2.2. Transmission and reflection of light from lossy dielectric layers.

$$M_j = \begin{bmatrix} \cos\Phi_j & \dfrac{i\cdot\sin\Phi_j}{\delta_j} \\ i\cdot\delta_j\cdot\sin\Phi_j & \cos\Phi_j \end{bmatrix}$$

[2.7]

where δ_j equals to;

$$\delta_j = \begin{cases} N_j\cdot\cos\theta_{tj} & \text{TE - polarization} \\[2mm] \dfrac{N_j}{\cos\theta_{tj}} & \text{TM - polarization} \\[2mm] N_j & \text{Normal - incident} \end{cases}$$

[2.8]

If there are h number of stacked dielectric slabs, then the transfer matrix representing light traveling from input layer (layer 0) to the output layer (layer h+1) is the multiplication of each individual layer's transfer matrixes as given in equation [2.9].

$$M_{TOTAL} = \prod_{j=1}^{h} M_j = \begin{bmatrix} m_{11} & m_{12} \\ m_{21} & m_{22} \end{bmatrix}$$

[2.9]

Amplitudes of transmission (t) and reflection (r) coefficients are defined in terms of transfer matrix elements with the following equations.

$$t = \frac{2\cdot\delta_0}{\delta_0\cdot m_{11} + \delta_0\cdot\delta_{h+1}\cdot m_{12} + m_{21} + \delta_{h+1}\cdot m_{22}}$$

[2.10]

$$r = \frac{\delta_0\cdot m_{11} + \delta_0\cdot\delta_{h+1}\cdot m_{12} - m_{21} - \delta_{h+1}\cdot m_{22}}{\delta_0\cdot m_{11} + \delta_0\cdot\delta_{h+1}\cdot m_{12} + m_{21} + \delta_{h+1}\cdot m_{22}}$$

[2.11]

Total transmitted and reflected power through multi layer dielectric slabs, or transmittance (T) and reflectance (R), are given by following equations.

$$T = \frac{\delta_{h+1}}{\delta_0}\cdot\left|t^2\right|$$

[2.12]

$$R = \left|r^2\right|$$

[2.13]

Transmittance is a measurement of how much of impinging light is transmitted to the next layer. If T equals one, than the material is none absorbing. If it equals zero than all impinging light energy is absorbed in the material or reflected. These equations are used for calculating layer thickness or dielectric layer properties to develop thin antireflective coatings for image sensors or materials.

2.1.2 Photon Conversion

After passing through multiple overlaying dielectric layers, impinging photons reach to semiconductor surface. Like overlaying isolation layers [Hu03], silicon (Si) has a complex refractive index as shown in Figure 2.3 [Groom99]. Extinction coefficient (k_j) of silicon is dominant below 300nm. Because of this, short wavelength photons are absorbed close to the surface of the semiconductor as shown in Figure 2.4 [Groom00].

Photons that have enough energy penetrate into the semiconductor substrate and transfer some portion of their energy to the semiconductor lattice. If this energy is high enough, electron(s) will be released from the valance band to conduction band leaving a hole behind (in p-type silicon substrate). Photon energy is given with equation [2.14].

$$E = \hbar \cdot \upsilon = \frac{\hbar \cdot c}{\lambda} = \frac{1.2395}{\lambda(\mu m)} \quad (eVolt) = \frac{1.985575 \cdot 10^{-19}}{\lambda(\mu m)} \quad (Joule) \qquad [2.14]$$

where λ is the wavelength of impinging photons, c is speed of light at free space, and $\hbar$ is Planck's constant ($\hbar$=6.6262E-34 Joule*second).

For silicon, energy gap between conduction and valance band is approximately 1.1 eVolt. This means that the photons with higher than 1.1 eVolt energy could form electron-hole pairs in silicon substrate. Thus, photons that have shorter than 1100 nm wavelength are eligible for silicon base imaging.

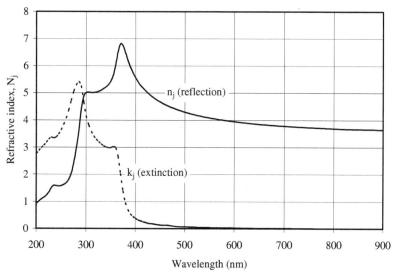

Figure 2.3. Reflection and extinction coefficients of silicon. (after [Groom99])

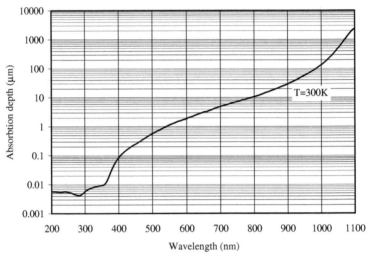

Figure 2.4. Absorption depth of light in silicon. (after [Groom00])

In dark environments, response curve of human eye shifts towards UV spectrum and its relative sensitivity increases. Human eye is sensitive enough to detect 214 and 126 photons per second for photons with 450nm and 650nm wavelength, respectively. At its peek point of 555nm, it can detect 10 photons per second. Human vision system is most sensitive to the light spectrum between 380 nm and 770 nm and the optical characteristics of silicon well cover this range. Thus, most of today's state-of-the-art image sensor systems are developed on silicon-based technologies.

2.1.3 Charge Separation and Collection

After absorption of photons, and generation of electron-hole pairs in semiconductor bulk material, negatively charged electrons have to be separated from positively charged holes. Easiest way to do this is to apply an electric field by which electrons are captured and holes are drained (or vice versa, depending on the type of semiconductor substrate material).

Captured electrons (or holes) could be counted to quantify amount of light dropped on photo sensitive conversion site. However, it becomes very difficult to count every generated electron when considering photonic and electronic noises. In current mode photo detectors, electrons are collected and counted continuously. Other way to count electrons (or holes) is to integrate them in a charge pocket and read the integrated charges in predetermined time intervals. After integrated charge has been read, charge pocket is reset to a known level, and charge integration starts again. This type of image sensors are known as integration type imagers and most modern image sensor technologies are based on this operation principle.

10

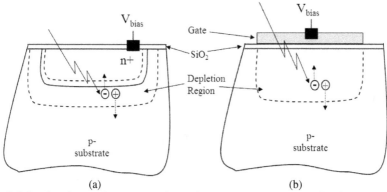

Figure 2.5. Semiconductor photon conversion and storage structures, a) metallurgical p-n junction, the photodiode, b) voltage-induced conversion and storage, the photogate.

Two alternative methods to form electric field in semiconductor material to separate electron-hole pairs and at the same time store them in a build-in capacitance are depicted in Figure 2.5. First one uses metallurgical p-n junction and the other uses voltage induced photon conversion and collection site. Electron-hole separation is managed by the electrical field across the depletion regions. Electrons are accumulated at the reverse biased side of the p-n junction (photodiode) or at the bottom plate (the channel) of the MOS capacitor that are electrically isolated from the other 3D structures or pixels. These structures are known as the photodiode and photogate.

2.1.3.1   *Photodiode (PD)*

Cross section of a p-n junction photodiode formed in an n-well CMOS process is depicted in Figure 2.6. Photodiode is reverse-biased and formed by using a shallow n+ doped, drain-source diffusion of an NMOS device, and the grounded p- type substrate. A bias voltage is applied to the n+ region to form a depletion region around the metallurgical p-n junction. This depletion region is free of any charge because of the electrical field. Any electron-hole pairs generated in this region sees the electrical field as shown in the A-A′ cross-sectional view of the photodiode in Figure 2.6. Electrons "slide" at the opposite direction of electric field towards the n+ region, while the holes go towards the p- region. Electrons are collected in the charge pocket in the n+ region while the holes are damped to the ground, or they are recombined. This type of photodiode is widely used in CMOS image sensor technologies and early CCDs as photo conversion and collection element.

There are issues with using standard n+ diffusion layer of NMOS transistor as photosensitive element. One is the dark current induced by the stress centers around the n+ diffusion. These stress centers form around extended field oxide (FOX) type separation walls between devices in standard sub-micron, self-aligned CMOS processes. Cross sectional view of

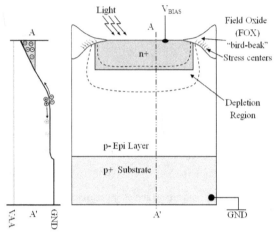

Figure 2.6. Cross-section of p-n photodiode, and associated potential-well diagram.

these extended regions on n+ diffusions looks like "bird beak" as shown in Figure 2.6. Another issue is surface related dark current and quantum loss in shorter wavelengths. This is because of the work function difference between the n+ diffusion surface and the overlaying isolation oxide layer. This causes surface recombination centers and defects. These localities absorb any photo generated electron-hole pairs close to the photodiode surface. Short wavelength photons, or blue photons, are absorbed on the surface of the silicon. Thus, photodiode is less sensitive in blue spectrum.

2.1.3.2   *Buried (Pinned) Photodiode (BPD)*

Buried or pinned photodiode type photon detectors have been successfully used in CCDs for many years [Teranishi82] [Burkey84]. Cross section and potential-well diagram of a buried photodiode in an n-well CMOS process is depicted in Figure 2.7. It is called "buried", or "pinned" because, charge storage region, or n- diffusion of the photodiode is buried below the silicon surface by a p+ region. Surface p+ region allows depletion region to extend all around the n- diffusion. Ideally, depletion region extends up to the surface leading to a better short wavelength photon collection, or blue response. Better blue response is one of the advantages of buried photodiodes. Another advantage is the reduction of dark current. This is because, the n- diffusion does not have surface and stress related dark current components.

One of the major disadvantages of buried photodiode is that they are not available in standard CMOS processes. Extra masks have to be used to form surface p+ and n- regions in standard CMOS processes. In addition, doping concentration of the n- photodiode and surface p+ regions must be optimized for better imaging performance. Thus, buried photodiodes require specialized CMOS process like CCD.

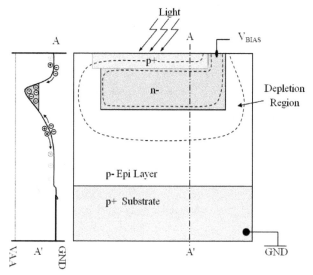

Figure 2.7. Cross section and potential-well diagram of the buried photodiode.

2.1.3.3 Photogate (PG)

Cross section and potential-well diagram of photogate type photosensitive element are depicted in Figure 2.8. Gate side of a MOS capacitor is biased to induce a depletion region underneath the gate in photogate. Photogate was first used in CCDs. Later, it was adapted in CMOS image sensors for low noise charge collection operation [Mendis93a, Fossum93b].

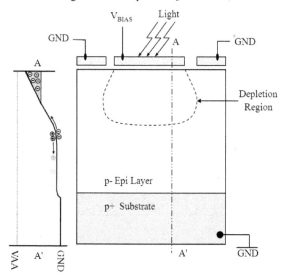

Figure 2.8. Cross section and potential-well diagram of the photogate.

Like photodiode, the charge pocket extends up to the surface of silicon. However, in photogate, the surface or the MOS device channel has better quality than that of the photodiode. Thus, the work function between channel and gate oxide is improved leading to lower surface dark current. In addition, there is no stress related dark current component in photogate. These lead to a very low dark current photo sensing operation. In photogate type photon converters, optical property of the MOS gate material is very important. Gate material has to be transparent to visible light and should have good optical properties to reduce loss or reflection of photons before reaching to the depletion region. In modern sub-micron CMOS technologies, poly-silicon gate material with special refractory materials deposited on top (gate-silicide or salicide process) are used to reduce gate delay. These refractory materials make the poly silicon gate virtually opaque to visible light. Hence, a blocking layer must be used which protects photogate poly silicon during the gate-silicide or salicide process step.

2.1.3.4   Buried Channel Photogate (BPG)

Similar to the buried photodiode photon converters, charge collection pocket could be buried below the silicon-MOS gate surface to further reduce surface related dark current in photogate type photon converters. Cross section and potential-well diagram of buried channel photogate on a p-type substrate is shown in Figure 2.9. In this case, pinning is achieved by a very thin n- doping layer deposited uniformly just underneath the gate oxide material.

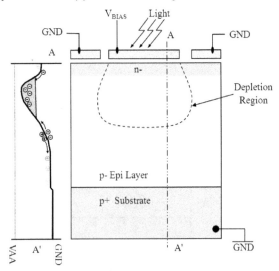

Figure 2.9. Cross section and potential-well diagram of the buried channel photogate.

2.1.4   Electron to Digital Data Conversion

After impinging photons are converted and collected in confined 3D structures, they are read out to higher processing circuits. These circuits either located on or off chip depending on the technology as depicted in Figure 2.10 for CCD and CMOS based image sensors. Major role of these circuits are to convert pixel charge pockets to measurable analog or digital information. In CMOS based image sensors, pixel signals are buffered in voltage or in current form to column readout circuits. Buffering can be done in the pixel site or in the column.

After buffering, pixel signal and pixel-offset values are subtracted to get the absolute pixel signal values. This is called correlated double sampling (CDS). This absolute pixel signal is typically send to a programmable gain circuitry that further amplifies the absolute value. After gain stage, extra processing is performed to adjust amplified pixel signal to the input range of analog to digital converter (ADC). ADC converts the analog input signal to an n-bit digital output data. This digital data then processed in digital domain and sent off-chip for further processing or interpretation by imaging system. In modern CMOS image sensors almost all of the steps up to this point are performed on-chip. In CCDs however, only the charge transfer, and buffering are performed on-chip, and rest is done by off-chip components.

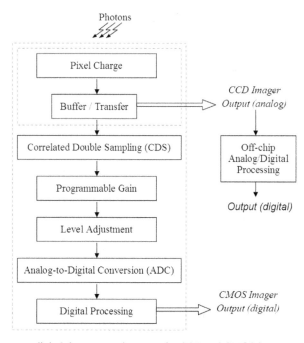

Figure 2.10. Electron to digital data conversion steps for CCD and CMOS image sensors.

2.2 Pixel Definitions and Performance Metrics

In this section, major definitions of photon conversion site called pixel and its performance metrics were defined to form a consistent terminology throughout this book.

2.2.1 Pixel

A pixel is an individual photo sensitive site that has the ability to collect photo generated charges, and to restrict the location of charges to a discrete volume of space within the silicon substrate. Photodiode and photogate are the two types of pixels that could be used to collect and to hold photo generated charges. CMOS active pixel sensor (APS) contains a preamplifier, reset and access transistors in addition to photosensitive region or element. Pixel size or pixel pitch is limited by technology parameters (design rules) of the fabrication process in use, and electrical and optical specifications of the product under development.

2.2.2 Floating Diffusion (FD)

Floating diffusion (FD) is a charge-sensing node used to convert charge packets (generated by the photo conversion site) into a change in voltage (or current) that could be buffered. In an integration type image sensor, floating diffusion node of pixel is reset to a pre-determined voltage before scene integration starts. It is left floating during charge integration period. When charge is dumped or collected onto pixel FD node, a proportional change in voltage occurs. The change in voltage due to a charge packet size of N_{SIG} (in number of electrons) on the charge-sensing node, FD, is given by the following simple equation.

$$V_{SIG} = \frac{q \cdot N_{SIG}}{C_{FD}}$$

[2.15]

where C_{FD} is the effective floating diffusion node capacitance in farad and q is the elementary charge (q=1.602*10^{-15}).

2.2.3 Fill Factor (FF)

Fill Factor (FF) or the pixel fill factor is the ratio of light sensitive area of photo conversion site (A_{PD}) in a pixel to total pixel square or the area (A_{PIXEL}). It is given in percent of pixel area with the following equation.

$$FF = \frac{A_{PD}}{A_{PIXEL}} \times 100 \quad [\%]$$

[2.16]

16

In some cases, both the photosensitive region and unblocked substrate areas around the photo conversion sites are considered. This is because of the fact that not only the charges converted in the photosensitive areas are collected but also the charges generated at the vicinity of the photosensitive areas and traveled laterally to the photo collection regions.

2.2.4 Conversion Gain (CG)

After the impinging photons are converted and collected at pixel floating diffusion nodes, collected electrons cause a proportional change in voltage depending on the FD node capacitance (C_{FD}). This is called charge-to-voltage conversion gain, or simply conversion gain (CG). It is given in microvolt per electron ($\mu V/e^-$) with the equation given in [2.17].

$$CG = \frac{V_{FD}}{N} = \frac{q}{C_{FD}} = \frac{160.2 \times 10^{-15}}{C_{FD}}. \qquad (\mu V/e^-) \qquad [2.17]$$

2.2.5 Full-Well Capacity

Maximum amount of charge that an imaging pixel could collect and transfer while maintaining all the performance parameters is called saturation charge level. This defines the pixel full-well capacity. Full-well charge capacity may be limited either by the size of the photo conversion site or by the size floating diffusion region, and by the signal readout circuit's ability to buffer pixel signals during readout.

Dividing the charge-to-voltage conversion gain (CG) with the pixel saturation voltage results in pixel full-well capacity in number of electrons (N_{SAT}) and it is given with the following equation.

$$N_{SAT} = \frac{V_{SAT}[\text{Volt}]}{CG\left[\frac{\mu Volt}{e^-}\right] \times 10^6} \qquad (e-) \qquad [2.18]$$

2.2.6 Quantum Efficiency (QE)

Ratio of the number of photo generated electrons captured by a pixel to the number of photons incident upon the pixel is called quantum efficiency (QE). Quantum efficiency is measured and defined at different wavelengths. Wavelength dependent quantum efficiency QE(λ) is given with the following equation.

17

$$QE(\lambda) = \frac{N_{SIG}}{N_{PH}(\lambda)}$$ [2.19]

where N_{SIG} and $N_{PH}(\lambda)$ are the generated signal charge and the number of incident photons in the pixel, respectively. λ is the wavelength of the photons.

Quantum efficiency varies with optical wavelength due to the wavelength dependence of the absorption depth of the semiconductor bulk materials, the transmission-reflection properties of the overlaying layers of the photosensitive site, and the fill factor of the pixel.

2.2.7 Dynamic Range (DR)

Dynamic range (DR) is defined as the ratio between the maximum signal, or saturation level that could be imaged by a pixel and the noise level in absence of light. Saturation and noise levels are typically referred to the pixel floating diffusion node in charge domain. In charge domain, input referred noise and signal levels could be calculated by dividing noise and signal output voltages with image sensor's measured conversion gain (CG, $\mu V/e^-$). Dynamic range usually is expressed in decibel (dB) with the following equation.

$$DR = 20 \cdot \log\left(\frac{N_{SAT}}{N_{NOISE\_FLOOR}}\right)$$ [2.20]

where N_{SAT} and N_{NOISE_FLOOR} are the saturation charge (full-well capacity) and the input referred noise floor in electrons, respectively.

2.2.8 Linear Dynamic Range (LDR)

Linearity of an image sensor is deteriorated at the region of near saturation due to nonlinearities on the signal readout circuitry. Linear dynamic range (LDR) is defined by the following equation.

$$LDR = 20 \cdot \log\left(\frac{N_{LINEAR\_MAX}}{N_{NOISE\_FLOOR}}\right)$$ [2.21]

where N_{LINEAR_MAX} is the maximum number of charge above which the linearity starts to degrade (the point at which linearity deviates about 1% from the straight line). Linear dynamic range is more important than saturation dynamic range as given in [2.20]. Linear dynamic range of a pixel typically relates saturation signal level to pixel full-well capacity. Photon transfer curve peaks at pixel saturation signal level.

2.2.9 Intrascene Dynamic Range (IDR)

Intrascene dynamic range (IDR) refers to the range of incident signal that can be accommodated by an image sensor in a single frame of captured image. Examples of scenes that generate high dynamic range incident signals include an indoor room with outdoor window, outdoor with mixed shadow and bright sunshine, night time scenes combining artificial lighting and shadows, and in automotive context, a vehicle entering or about to leave a tunnel or shadowed area on a bright day.

To cover a single scene that might involve indoor lighting (100 Lux) and outdoor lighting (50,000 Lux), required intrascene dynamic range is of the order of 5000:1 (assuming 10 Lux-equivalent noise) corresponding to 74 dB. In digital bits, this requires 13-14 bits of resolution on 1 volt signal swing. An image sensor may have a larger IDR if the exposure time is to be adjusted for different light levels for individual pixel or pixel segments.

2.2.10 Responsivity

Responsivity is defined as the ratio of the optical detector's output photocurrent (or voltage) in amperes (or volts) to the incident optical power in watts. Responsivity ($R(\lambda)$) depends on wavelength of the incident light, quantum efficiency, photo collection area, and the incident light intensity. It can be expressed with the following equation.

$$R(\lambda) = \frac{I_{curr}}{P_{in}} = \frac{q \cdot QE(\lambda) \cdot \lambda}{\hbar \cdot c} \qquad \text{(A/W)} \qquad\qquad [2.22]$$

where λ is the wavelength of the photon, c is the speed of light, and the $\hbar$ is the Planck's constant.

2.2.11 Sensitivity

Sensitivity is the ratio of the pixel output change (voltage or current) obtained after an exposure time to the amount of light change that has a specific wavelength. Typically, amount of light change during the exposure time is given in specific wavelength in Lux∗seconds. Sensitivity is given either in volts per Lux∗seconds or in electrons per Lux∗seconds, or in bits per Lux∗seconds. During the measurement, voltage on the pixel floating diffusion is plotted against the different amounts of light exposures. Sensitivity is then the slope of the line between the dark current offset at zero Lux∗second and the saturation point. Saturation point generally set at the deviation point of a curve from a straight line within 1% linearity. Digital sensitivity measure is used in case the image sensor has an on-chip ADC. The value becomes larger with an image sensor having higher ADC resolution, even though the 'analog' sensitivity is the same.

19

2.2.12 Signal-to-Noise Ratio (SNR)

Signal-to-noise ratio (SNR) is the ratio of the input referred signal level (N_{SIGNAL}) to the input referred noise level (N_{NOISE}) at certain illumination level in decibel. It is given with the following equation.

$$SNR = 20 \cdot \log\left(\frac{N_{SIGNAL}}{N_{NOISE}}\right) \qquad [2.23]$$

Photon shot noise is the dominant noise source in high illumination levels, while read noise (noise floor) is dominant in low light levels. At low exposure levels where the signal is lower than the noise floor, SNR increases with exposure by 20 dB per decade. It increases by 10 dB per decade at high exposure levels where the photon shot noise is dominant as seen in the Figure 2.11.

The highest achievable SNR (peek SNR) is typically reported as the SNR level and calculated using the equation [2.24]. As seen from the expression, the peek SNR only depends on the saturation charge N_{SAT} (full well charge).

$$(SNR)_{MAX} = 20 \cdot \log\left(\frac{N_{SAT}}{N_{NOISE\_SHOT}}\right) = 20 \cdot \log\left(\frac{N_{SAT}}{\sqrt{N_{SAT}}}\right) = 10 \cdot \log \cdot N_{SAT} \qquad [2.24]$$

Discrimination threshold of human vision is approximately a constant ratio of luminance over a range of intensities of about 300:1. It means that human visual system cannot recognize difference in intensity less than 2% of average luminance levels. Considering this difference as peak-to-peak range, we can calculate the rms value of acceptable noise as 0.65% with accuracy almost 90%. Therefore, a high quality camera needs the signal-to-noise ratio at least 150 or 44 decibel (dB).

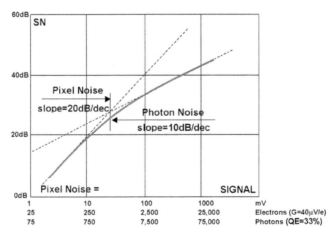

Figure 2.11. Signal-to-noise ratio (SNR) versus signal.

2.2.13 Noise Equivalent Exposure (NEE)

Noise Equivalent Exposure (NEE) may be a good measure of sensitivity. It is straightforward when comparing image sensor sensitivities. NEE is defined by the exposure in Lux∗seconds at which the signal is equal to noise. In specifying NEE, it is important to address the integration time and temperature since the dark signal is a function of integration time and temperature, while the signal charge is a function of exposure. In addition, description of the light source and an IR cut filter is needed.

2.2.14 Blooming / Smear

If pixel full-well charge capacity is exceeded, the excess charge will overflow into adjacent structures and produce artifacts known as blooming and smear. In CCDs, if the charge in a over saturated pixel leaks out into the next pixel it cause blooming, while if it leaks into the charge transfer channel it cause a white streak on the image called smear. Typically, an antiblooming (AB) structure is placed nearby the charge collection site preventing excess charge to overflow into adjacent charge sensitive structures.

2.2.15 Dark Current

Dark current is referred to the background signal present in image sensor readout channel when no light is incident upon the image sensor. This background signal is the result of thermally emitted charges which are being collected in photosensitive region of pixel. Usually, dark current is quoted in pico ampere per square area ($pA*cm^{-2}$). It dependents on fabrication process, sensor architecture, operation mode, and temperature. Dark current doubles for approximately every 5 to 10 degree (C) increase in temperature. In general dark current is given with the following equation.

$$I_{DARK} = \frac{q \times N_{DARK}(T)}{t_{INT}}$$
[2.25]

Three major contributors of dark current are; generation of electrons in surface states at Si-SiO_2 interfaces, generation of electrons in depletion regions in generation-recombination centers of silicon, and thermal generation of charges in neutral bulk and diffusion of those charges to potential wells. Intrapixel dark current could be generated at different pixel locations due to irregularities in pixel structures. Any interpixel dark current nonuniformity would cause pixel-wise fixed pattern noise (FPN). This could only be removed by complicated signal processing techniques [Theuwissen95].

21

2.3 Noise Sources in CMOS APS

Like any other electronic circuit, image sensors not only produce output signals, but also generate different kind of noises. There are two major groups of noise recognized in CMOS APS sensors: Fixed pattern noise (FPN) and temporal noise.

2.3.1 Fixed Pattern Noise (FPN)

If image sensor output is viewed under fixed uniform illumination, a fixed, non-temporal, spatial pattern can be seen on the image. This is called fixed pattern noise (FPN). It refers to pixel-to-pixel, row-to-row, or column-to-column variations of the video output images. FPN appears due to the variations of individual pixel parameters. These variations occur in the optical path before photoelectric conversion and in the electrical path after conversion has taken place. FPN could be specified either as a peak-to-peak value or root mean squared (rms) value referenced to the signal mean.

Originally, CMOS APS pixels had a very serious problem with two-dimension FPN due to threshold voltage variations of MOSFET transistors exist in each pixel. In each pixel, these variations accounted for 10 to 30 mVolt rms non-uniformity. In addition, if no special correction was applied, FPN in the column readout circuits reach 3-5mVolt rms. Differential delta sampling (DDS) practically eliminates both of these problems [Nixon96].

It is convenient to divide FPN into two principally different components based on signal dependence. One is the signal independent FPN or dark signal non-uniformity (DSNU), and the other one is the signal dependent FPN or photo response non-uniformity (PRNU).

2.3.1.1   *Dark Signal Non-Uniformity (DSNU))*

Pixel-to-pixel dark current variation is one of the sources of two-dimension FPN. This variation is not signal dependent and called dark signal non-uniformity (DSNU). It is due to non-uniform spatial patterns of impurity concentrations in the wafer. Localized concentrations of donor impurities that cause cloud like patterns on video image are another cause of the DSNU. Additionally, this non-uniformity depends on temperature distribution on the pixel array area.

Dark voltage average (V_{DRK}) refers to average output value in the absence of incident light. Dark signal non-uniformity (DRNU) refers to distribution of dark output voltages for each pixel as depicted in Figure 2.12. Dark voltage average (V_{DRK}) is caused by charge generated without any relationship to light. Because, this charge is generated on a steady state condition, and dark voltage average is proportional to the storage time.

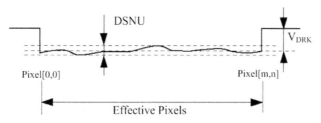

Figure 2.12. Dark voltage average (V_{DRK}) and dark signal non-uniformity (DSNU)

2.3.1.2 *Photo Response Non-Uniformity (PRNU))*

Photo response non-uniformity (PRNU) is the signal dependent component of FPN. Local variations in different layer thicknesses and doping impurities cause variations in photogenerated carrier lifetime. These are caused by photolithographic irregularities such as mask misalignment. They result in modifications in QE, pixel capacitance, or source follower gain across the photosensitive array. PRNU depends on process technology, light spectrum, pixel design, operation mode, and timing. PRNU refers to situations where output signal amplitude varies for each pixel in the array when they are exposed to a uniform illumination as shown in the Figure 2.13. Typically, PRNU defined with the following equation.

$$PRNU = \frac{(V_{MAX} - V_{MIN})}{2 \cdot V_{AVE}} \times 100 \qquad (\%) \qquad [2.26]$$

V_{MAX} is the maximum output of effective pixels, V_{MIN} is the minimum output and V_{AVE} is the average output. For color image sensors, PRNU is defined for each color.

In general, if ADC has enough accuracy, FPN can be removed completely. DSNU is relatively easier to correct than PRNU. Dark frame or row subtraction will reduce the two-dimensional or vertical component of FPN, respectively.

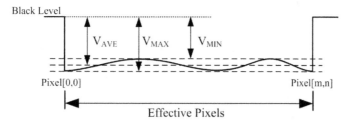

Figure 2.13. Definition of photo response non uniformity (PRNU)

2.3.2 Temporal Noise

Temporal noise refers to time-dependent fluctuations in signal levels that are fundamentally different from the FPN. In general, temporal noise can be divided in three category; shot noise, 1/f noise, and thermal noise. In a solid-state image sensor, several noise sources contribute to temporal noise. These are pixel photodiode's shot noise, pixel reset and readout transistor's thermal, shot, and 1/f noises, and column amplifier's thermal and 1/f noises. Reset and readout transistor's thermal, 1/f, and shot noises are the dominant temporal noise source in low illumination. While at high illumination, dominant noise source is the photodiode's photon shot noise.

2.3.2.1 Photon Shot Noise

Photon shot noise, otherwise known as Schottky noise, is a broadband fundamental noise that stems from counting statistics. A flow of electrons or photons, rather being continuous, is composed of a stream of discrete particles. When number of particles that passes a given point must be determined, Poisson statistics will govern the average number of particles that pass per unit time, and the error in that number. Error associated with the number of particles that have passed is given by

$$\sigma = \sqrt{N} \tag{2.27}$$

where σ represents the standard deviation, and N is the number of particles that have been counted. For systems in which N is large, standard deviation is very small in comparison with N. For very low signals levels, however, value of N becomes small, and so does the signal to noise ratio (SNR= N/σ).

Probability of a detector converting a photon into a unit of charge is defined by the quantum efficiency. Quantum efficiency determines what portion of a population of photons striking a detector will be detected. Moving charges produced within the detector constitute a current. It's level determined by the detector's sensitivity (A/W/cm^2). If the value of this current is high enough and no signal-dependent excess noise is generated within the detector, shot noise from the photon flux will dominate the noise for the system. As can be seen from the equation [2.28], amount of noise seen will be proportional to the square root of the number of photons. Amount of noise will also be proportional to the square root of the detection bandwidth, because as bandwidth is decreased, more photons are counted concurrently.

$$n = \sqrt{N_{ph}} \tag{2.28}$$

2.3.2.2   Flicker (1/f) Noise

Low frequency noise in MOSFET is dominated by flicker noise. It is commonly known as 1/f noise, because, the noise spectral density is inversely proportional to the frequency of signal. It is also inversely proportional to the MOSFET's gate area. The effect of 1/f noise on sub-micron, low frequency CMOS analog circuits, such as CMOS APS image sensor electronics, becomes more and more pronounced.

In general, 1/f noise originates from fluctuation in the conductivity. This could be caused by fluctuations in the mobility, or in the total number of charge carriers, or both [ChangJ94, Hung90]. In CMOS APS pixel, different components contribute to total 1/f noise at different operation phases. For example during integration time, 1/f noise is produced by the photodiode dark current fluctuation. It is generally much smaller than the photodiode shot noise and thus is being ignored. During readout phase, the column source follower and access transistors generate 1/f noise. Fluctuation in the drain current of these transistors can be written as [Tian00];

$$S_{Id}(f) = g_m^2 \cdot S_{Vg}(f) = g_m^2 \cdot \frac{1}{C_{ox}^2} \cdot S_{Qch}(f) = g_m^2 \cdot \frac{t_{ox}^2}{\varepsilon_{ox}^2} \cdot \left[\frac{q}{A}\right]^2 \cdot \left(\frac{kT \cdot A \cdot N_t}{2 \cdot \gamma \cdot f}\right) \qquad [2.29]$$

where kT is the thermal energy, A is the MOSFET gate area, t_{ox} is the effective gate oxide thickness, N_t is the gate oxide trap density (ev^{-1}cm^{-3}), γ is a constant, f is the frequency, g_m is the transconductance. $S_{Vg}(f)$ is the equivalent gate voltage 1/f noise power spectral density (PSD), and $S_{Qch}(f)$ is the channel charge density 1/f noise PSD. In SPICE circuit simulators, $S_{Vg}(f)$ is modeled as [Cheng98];

$$S_{Vg}(f) = \frac{k_F \cdot t_{ox}}{2 \cdot \varepsilon_{ox} \cdot A \cdot f} \qquad [2.30]$$

where k_F is a model parameter given in the SPICE model card. Equation [2.29] can also be written as equation [2.31] substituting g_m.

$$S_{Id}(f) = \frac{\mu_n \cdot t_{ox} \cdot q^2 \cdot kT \cdot N_t}{L^2 \cdot \varepsilon_{ox} \cdot f} \cdot i_{ds} \qquad [2.31]$$

From designer perspective, two parameters are used for reducing 1/f noise; channel length (L) and bias current (i_{ds}). Channel mobility (μ_n) and the oxide thickness (t_{ox}) are the two parameters controlled and optimized by the process engineers. In system level, reducing the system temperature (T) helps improving the 1/f noise. Correlated double sampling (CDS) used in CCD and CMOS image sensors does not necessarily suppress 1/f noise.

25

Reset or kTC noise is caused by uncertainty of the voltage on floating diffusion (FD) capacitance (C_{FD}) after pixel reset operation. This uncertainty could be quantified as number of charges. It is defined in equation [2.32] for "hard" reset mode of operation for CMOS APS pixels.

$$Q_{kTC} = \sqrt{k \cdot T \cdot C_{FD}}$$ [2.32]

Standard "hard" reset is the operation mode in where voltage on the drain of the reset transistor has smaller than that of its gate during reset period. k here represents Boltzmann constant and T is the temperature in Kelvin. Typically, "soft" reset mode of reset operation is used in CMOS APS pixels. In this case drain and gate voltages of the reset transistor are set to same level during reset period. Thus, reset transistor operates in incomplete charge transfer mode. Because discharge process becomes emission limited, uncertainty of reset level on FD drops to [Thornber74];

$$Q_{kTC}^* = \sqrt{k \cdot T \cdot C_{FD}/2}$$ [2.33]

2.4 Correlated Double Sampling (CDS)

A sampling method called correlated double sampling (CDS) is utilized in solid-state image sensor to eliminate temporal noise from image sensors [White74]. In CDS, voltage of the pixel floating diffusion is read twice; just after the pixel reset and after the signal integration period ends. Difference between these two samples is the useful signal with the reset noise eliminated.

In a solid-state image sensor, photo generated charge is typically collected on a capacitor. Voltage across this capacitor is read as the signal amplitude (V_S). During CDS operation the signal voltage ($V_S = Q_S/ C_{PIX}$) is compared with the "dark" or "reset" level voltage ($V_R = Q_R/ C_{PIX}$) that is obtained when all charges of the capacitor (C_{PIX}) have been reset to a fixed potential. Thus, for each pixel, an absolute pixel voltage can be determined by taking the difference of these two signals.

$$V_{PIX} = V_S - V_R = \frac{Q_S - Q_R}{C_{PIX}}$$ [2.34]

CDS cannot remove second order effects due to pixel gain non-uniformity, and cannot remove uncorrelated temporal white noise originating from before the signal subtraction operation, such as broadband amplifier noise. None of the downstream noise sources, such as EMI, system noise, etc. is affected. Low frequency MOSFET noises (1/f noise, flicker noise) are reduced only partially. Signal noise, such as photon shot noise, does not effected by CDS.

True CDS operation can be performed in a pinned-photodiode and photogate type APS pixels. Because reset and charge integration nodes are isolated. In simple 3T photodiode type APS pixel, on the other hand, charge integration and reset node are the same. Therefore, it is not possible to perform a true correlated double sampling operation during the same frame period.

2.5 Optical Limitation

In the development of a large CMOS image sensor array, numerous optical limitations must be considered. These limitations include fill factor, pixel sensitivity, and crosstalk. These limitations and other parameters are discussed in this subsection.

2.5.1 Imaging Optics

To understand the whole imaging system, it is necessary to understand both the image sensor and the imaging optics. One fundamental parameter of the imaging optics (i.e., a lens) is its focal length, f. A lens' focal length (f) is the distance from the center of the lens that the lens will focus to a spotlight coming in from infinity. Second to the lens' focal length is the diameter of its aperture, A (Figure 2.14.a). A common parameter defining a lens' performance is its f-number, f/#. f-number of a lens is given approximately with the following equation.

$$f/\# \approx \frac{f}{A} \qquad \text{(Lens' f-number)} \qquad [2.35]$$

f/# determines the theoretical limit of how much light gets to the image plane and how sharply focused that light could become. In terms of the f/#, the lens' resolution (minimum theoretical spot size diameter, a) can be given approximately (for visible light) by the equation for the Airy disk as, (Figure 2.14.b):

$$a \approx (1.35)\, f/\# \qquad \text{(Minimum Spot Diameter)} \qquad [2.36]$$

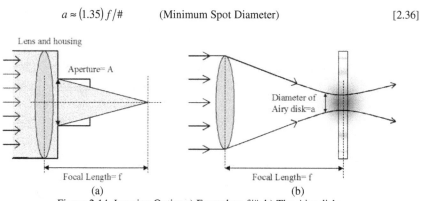

Figure 2.14. Imaging Optics a) F-number, f/#, b) The Airy disk

The diagonal full field of view (FOV) captured by a lens plus image sensor combination is given approximately as, (Figure 2.15.):

$$\theta_d \approx 2\tan^{-1}\left(\frac{d}{2\cdot f}\right) \approx 2\tan^{-1}\left(\frac{D}{2\cdot S}\right) \quad \text{(diagonal full field of view)} \quad [2.37]$$

In equation, $\tan^{-1}$ is the trigonometric function arctangent and d is the image sensor's diagonal dimension. In comparing different imaging systems, it is important that parameters such as field of view do not change.

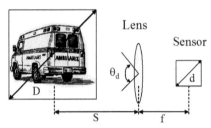

Figure 2.15. The diagonal full field of view captured by a lens and image sensor.

2.5.2 Light Budget

One critical parameter in an imaging system is the number of photons, p, that reach a single pixel during some given exposure interval. To calculate p, it is necessary to first calculate the illuminance at the sensor plane that a particular imaging situation produces. The illuminance (measured in Lux) at the image plane, $E_{v,i}$, is given by the following equation.

$$E_{v,i} \approx \left(\frac{R\cdot T\cdot \pi\cdot E_{v,o}}{4\cdot \beta\cdot (f/\#)^2}\right) \quad \text{(image plane illuminance)} \quad [2.38]$$

In equation [2.38], R is the power reflectivity of the object (0 – 1), T is the (f-number independent) transmittance of the lens (~ 1), $E_{v,o}$ is the illumination onto the object (in Lux), and β is the solid angle (in steradians) that the bulk of the light incident on the object scatters into (0 – 4π). From equation [2.38], it is then possible to calculate number of photons (p) through the following equation.

$$p \approx (1000\cdot E_{v,o})\cdot \left(\frac{R\cdot T\cdot \pi\cdot t_{IT}}{\beta}\right)\cdot \left(\frac{z}{f/\#}\right)^2 \quad \text{(photons per pixel)} \quad [2.39]$$

In equation [2.39], t_{IT} is the integration time. From equation [2.39], it is obvious that doubling pixel pitch (z), quadruples the number of photons (p), as does halving the f/#. This is a

28

critical design point. It raises the fundamental questions of what pixel size gives the best overall performance at the most reasonable cost.

2.5.3 Photon Sensitivity

Once a given sensor array design is feasible within a given fabrication process, resulting light sensitivity of the designed pixels must be considered. While smaller pixels offer smaller array sizes, they also offer smaller sensitivities to light (even for identical pixel fill factors). It can be shown that the amount of light entering the cross-section of a pixel is given with the following equation.

$$\left(\frac{p}{I_o \cdot t_{IT}}\right) = k \cdot \left(\frac{R}{\beta}\right) \cdot \left(\frac{z}{f/\#}\right)^2 \quad \text{(photons per lux*second)} \qquad [2.40]$$

In equation [2.40], p is the number of photons per pixel, I_o is the illumination onto the object, k is a constant of proportionality, R is the reflectivity of the object, β is the solid angle the light scatters off of the object into, z is the pixel pitch. The left-hand side of equation [2.40] represents the number of photons per lux*second that enters pixel's cross-section. After photon absorption, electron collection, and signal conversion, this term becomes proportional to the number of bits per lux*second representative of the sensitivity of the image sensor array. For a static imaging condition, equation [2.40] indicates how smaller pixel pitches (z) will result in lower sensitivity (independent of pixel fill factor). Compounding the problem of low sensitivity in small pixels is the low pixel fill factor (FF) achieved when attempting to push the limits on pixel design. This is due to the occlusion, by pixel circuit elements, of the light attempting to pass through the layers of the pixel to reach its photosensitive area. When both pixel pitch and pixel fill factor are considered, the relative sensitivity (S) that can be generated by a pixel will be related by the following equation.

$$S \propto FF \cdot z^2 \quad \text{(relative sensitivity)} \qquad [2.41]$$

2.5.4 Crosstalk

Numerous signal-to-noise limiting effects can occur in image sensor arrays. One such critical parameter to take note of when considering the design of high performance image sensors is the electrical and optical crosstalk that would occur when the light intended for one pixel gets detected by another pixel as pixel pitches decrease. Electrical crosstalk comes from carriers that are generated deep in the semiconductor and diffused into neighboring pixels. Optical crosstalk comes from multiple reflections of the impinging photons in the upper structures of the pixel.

For gray scale or panchromatic image sensors, whose outputs will be displayed on monochrome monitors, and color image sensors, there will exist two forms of photonic crosstalk and two forms of electronic crosstalk. The first form of photonic crosstalk is due to the limited performance of the imaging lens to produce image detail as small as the pixel. This is an especially difficult problem when considering large image sensor arrays, such as 4Kx4K pixels, and small pixel pitches, such as 1.0 μm. The second form of photonic crosstalk is due to the cross-sectional area enclosed by the light cone that is focused into a single pixel, measured at the effective absorption depth below the pixel. This can occur in both panchromatic and color image sensors within the silicon layer itself, where finite absorption depths (0.5–3.0μm for the visible) allow photons to penetrate from the region of one pixel into the region of another pixel. After light is focused onto a pixel, it rapidly defocuses below the pixel in where it continues to be absorbed by the substrate. For example, light entering one pixel may spread under neighboring pixels before becoming absorbed, therefore being registered as part of the neighboring pixels' signals. Once the photons have been absorbed, the photoelectrons that they liberated must diffuse into the collection region of the pixels. The crosstalk that occurs as electrons diffuse from their point of liberation to their point of collection is one form of analog electronic crosstalk. While electron diffusion lengths in silicon are several tens of microns, the production of epitaxial layers, doped collection wells, and potentials in the pixel design generally work well to minimize this form of crosstalk. However, when pixel pitches become sufficiently small, diffusion-induced crosstalk can become significant.

Another form of analog electronic crosstalk occurs as pixel signals influence one another. Digital electronic crosstalk occurs as the digital signals propagating throughout the sensor array radiate into analog signal chains. With proper electronic design and placement, these latter two forms of electronic crosstalk can often be eliminated.

2.5.4.2 Above-the-Silicon Crosstalk in Color Image Sensor

In color image sensors, optical crosstalk can also occur in the layers between the color filter arrays (CFA) and the photo detectors. This above-the-silicon optical crosstalk is generally more serious, because the layers above the silicon have lower refractive indices than the silicon (~ 1.5 above the silicon, ~ 4 in the silicon) which allows the deviant light to propagate at a greater angle of incidence. These layers are also taller than the absorption depths in the silicon (3 – 10 μm tall typically). In an effort to reduce optical crosstalk, it is advisable to use as few metal layers as possible in the pixel circuitry design and/or keep the thickness of any passivation or planarization layers as small as possible. Perhaps the easier of these two suggestions would be to use, for example, two metal layers instead of three metal layers. Because, each metal layer adds approximately 1.5 µm of height to the pixel in today's CMOS technology. Thus, such an approach will produce a moderate reduction in the optical crosstalk for most pixels.

2.6 Image Sensor Characterization

Image sensor testing and characterization is one of the tedious parts of the design process. It requires multi level of hardware and software co-design to extract the image sensor characteristics with minimum added noise.

2.6.1 Environmental and Illumination Control

Image sensors are tested optically and functionally on automated or manual characterization setups in controlled environments. Environmental parameters are the room temperature, image sensor die temperature, humidity, background light, and stress.

The pixel array illumination is provided by on optical system in where a number of optical parameters are controlled. Typical parameters are intensity, uniformity, color temperature, and pattern of the light. Intensity of light on the image sensor's focal plane should be variable over at least four decades of dynamic range. Changeable spectral filters could be used to define spectral characteristics of the illumination. The color temperature of light source should be variable (or can be simulated by using color shift filters) in the range from 2600 K to 6500 K. Intensity of light on the focal plane could be measured by using embedded calibrated photodiode. Optical system should provide uniform illumination with spatial non-uniformity no more than that of the imaging pixels. Optical system could also produce images from test charts for measuring spatial characteristics of the photosensitive array. Testing setup should have interfaces with personal computer (PC) that controls these devices and processes image data received from the image sensor.

2.6.2 Characterization Board Testing

Characterization typically starts with measuring and optimizing noise sources of external support circuitry without image sensor chip (DUT - Device Under Test) plugged on the test board. First, normal biasing and clocks are applied to the test board and supporting circuitry without DUT plugged in. Then, the random readings reflecting noise induced by the test board (due to clocks and biasing) and supporting circuitry are monitored. The result with no input signal may not be a completely valid noise concern, but should provide information useful to the next step when real signals are used. Next step is characterization of the setup and noise with dc signal inputs. Still without DUT, DC signals (with noise less than the quantization noise of the ADC being used on the

31

system) are applied to the input paths of the ADC with various levels of voltages. In analog output image sensors, noise of the setup at simulated reset and signal levels are measured. It is done by applying an AC signal to the different input paths of the ADC. With clocks and biases monitored, the output represents channel specific noise and cross talk due to test board and test board components. Typically, first measurement with the DUT plugged on the test board is the system noise measurement. It is done by measuring output noise of various DUT's in dark condition at various integration times and with multiple insertions.

2.6.3 Analog Signal Chain Calibration

A calibration procedure is applied to the image sensor in advance to estimate correct signal values at the pixel output. During the calibration procedure, DC voltage generated by the testing system is applied to the analog signal chain (ASC) input instead of signals from the photosensitive array. As a result of this calibration the transaction factor from pixel output to ADC output (in LSB per Volt) is calculated for different ASC gain settings and light levels. These numbers are used to refer measured output signals to the pixel's floating diffusion node.

2.6.4 Sensitivity and Saturation Measurements

For sensitivity and saturation signal measurements, photosensitive pixel array is illuminated by uniform light through the Green or IR cut filter. The measurements start from the setting of an image sensor to the minimum integration time (usually this is 1 row of integration time). A number of frames (30 ÷ 40 frames) are acquired from the image sensor. The mean and standard deviation of pixel responses are calculated. Then the measurements are repeated for different integration times to receive light-signal characteristic (signal vs. exposure). Exposure is calculated as an illuminance multiplied by an integration time:

$$E = I \times \alpha \times t_{int} \qquad [Lux*sec] \qquad [2.42]$$

where I is illuminance in footcandle, [fc], α is wavelength dependent illumination conversion factor (equals to 10.764 for 550nm, for example), and t_{int} is the integration time in second.

All measurements are performed by increasing the integration time from minimum to the maximum value, at which signal from the image sensor is saturated. Signals from the image sensor are calculated both in LSB and in Volts at the pixel output. The sensitivity is calculated as a slope of the linear part of the light-signal characteristic. Usually, sensitivity is calculated at the pixel output in volt per Lux second [Volt/Lux-sec]. The sensitivity at the ADC output in LSB per Lux second [LSB/Lux-sec] at specified gain setting could be calculated as well.

2.6.5 Dark Current Measurement

Dark current can be specified as a number of input referred electrons generated per second in a pixel, or as a current per unit area. It is measured by plotting average pixel output voltages at different integration times while the image sensor is placed in a dark and controlled temperature environment. The slope of the signal versus integration time plot is used to calculate the dark current in volt per second (V/sec). If the conversion gain, fill factor, and pixel pitch of the image sensor is known, dark current is converted to ampere per squared area per second with the following equation.

$$J_{dark} = \frac{1.6 * 10^{-19} [C/e^-] \times V_{dark} [V/s]}{CG \ [V/e^-] \times A_{pixel} [cm^2]} \qquad [A/cm^2] \qquad\qquad [2.43]$$

where A_{pixel} is the pixel photosensitive area and CG is the pixels' conversion gain, and V_{DARK} is the measured slope of the signal versus integration time plot.

2.6.6 Conversion Gain Measurement

One of the most important parameter regarding the characterization of image sensors is the determination of the signal generated per photo electron known as the conversion gain (CG) and is defined as [Beecken96];

$$CG = \frac{\partial x}{\partial (QE \cdot \Phi)} \quad [\mu Volt/e^-] \qquad\qquad [2.44]$$

where QE is the quantum efficiency (photoelectrons per incident photon), Φ is the number of incident photons during the detector's integration period, and x is the detector's signal in appropriate units (e.g., mVolt). Using the Poisson statistics this equation can be written as;

$$CG = \frac{\sigma^2}{\overline{x}} \quad [\mu Volt/e^-] \qquad\qquad [2.45]$$

where σ is the variance of the detector's signal, and $\overline{x}$ is the mean value of the detector's signal. The trace of σ^2 versus $\overline{x}$ is called photon transfer curve. Measurement is done by stepped the light source from complete darkness to full well illumination in precisely measured increments. At each illumination level, typically 30 to 60 frames are captured and the mean and variance are computed. The exact illumination level for each measurement is recorded with a calibrated photodiode. Conversion gain is then computed by getting the slope of the photon transfer curve in the photon shot noise limited region. Variance and the mean values are calculated by using the following formulas [Ay02].

$$\overline{x} = \frac{1}{T} \cdot \sum_{k=0}^{T} \overline{x}_k \quad [\text{Volt}] \qquad (\text{Mean}) \qquad\qquad [2.46]$$

$$\sigma^2 = \frac{1}{M \cdot N} \cdot \sum_{i=0}^{M} \sum_{j=0}^{N} \sigma^2(i,j) \quad [\text{Volt}^2] \qquad (\text{Variance}) \qquad [2.47]$$

where $\overline{x}_k$, and $\sigma^2(i,j)$ are

$$\overline{x}_k = \frac{1}{M \cdot N} \cdot \sum_{i=0}^{M} \sum_{j=0}^{N} x_k(i,j) \quad [\text{Volt}] \qquad\qquad [2.48]$$

$$\sigma^2(i,j) = \frac{1}{T} \cdot \sum_{k=0}^{T} \left(x_k(i,j) - \overline{x}_k\right)^2 \quad [\text{Volt}^2] \qquad [2.49]$$

where i and j are the coordinates of a pixel, T is the number of frames that were collected, $\overline{x}_k$ and is the mean signal value of the frame k. $\sigma^2(i,j)$ is the variances of a pixel over T frames, M and N are the width and height of the calculation window.

2.6.7 FPN Measurement

FPN measurement performed at dark and 50% saturation of the image sensor. At the appropriate light level, multiple frames are captured and stored. A mean output signal value for the entire array is calculated. Then, the rms difference between individual pixels to the mean is divided by the mean output saturation signal and multiplied by 100%. FPN is typically expected to be lower than the human eye's perception level of 0.5% of the saturation level.

2.6.8 PRNU Measurement

Photo response non-uniformity (PRNU) is measured at 50% and 5% of saturation levels. At these light levels, one frame grabbed and stored. PRNU is calculated by dividing the standard deviation by the mean of the frame and multiplying by 100%.

2.6.9 Linearity and Dynamic Range Measurements

Linearity and the dynamic range are measured by grabbing frames at zero or shortest integration time, and grabbing multiple frames at certain integration time and at different light levels. Measured signal level is plotted against integration time multiplied by the light level. Dynamic ranges are calculated for different percent of linearity error bands, such as 0.5%, 1%, 1.5%, 2%, and 3%.

2.6.10 Quantum Efficiency Measurements

QE measurement is done by measuring the relative responsivity of the image sensor at different wavelengths. Quantum efficiency is measured in an imaging environment that is well defined and controlled environmentally. Typically, a very stable light source, a monochromator, and a calibrated photodiode are used in the QE measurement setup. The light source stability and controllability is very important to make sure that the image sensor and the reference, calibrated photodiode receives the same number of photons at adjusted spectrum. Monochromator provides a very sharp filtering of the light source before the image sensor. They form the spectrometer.

QE measurement starts with setting the spectrometer to an output wavelength and light level. Than the number of photons per pixel is measured and calculated by using the calibrated photodiode that is placed at exact location of the image sensor' face-plate. This process repeated for all available wavelengths. Than the image sensor is placed in the characterization setup and all the measurements are repeated. Measured output voltages from image sensor are converted into number of electrons detected by using the conversion gain. Then, QE is simply the number of detected electrons divided by the number of photons arrived per pixel.

If the light falling on the sensor (i.e., the "faceplate irradiance") is known to be $\varphi(\lambda)$ *photons/sec/cm²* (ascertained using a calibrated photodetector), then the quantum efficiency is equated as

$$QE(\lambda) = \frac{V_{pix}}{\varphi(\lambda) \cdot T \cdot A_{pix} \cdot G_{conv}}$$

[2.50]

In practice, the QE is averaged over a small neighborhood of pixels. Note that the formula used to calculate the QE applies to the cross-sectional area of an entire pixel. Dividing this area by the pixels' actual fill factor can yield larger numbers but is not generally performed since more than the actively defined region is photoresponsive [Photobit96].

2.7 Summary

In this chapter, basic definitions, principles, characteristics, and measurement methods of solid-state imaging were discussed. Solid state imaging process starts in the solid-state imaging element called pixel. Scene image, or the photons are converted into meaningful signals, either voltage or current, in pixels. Photons first pass through overlaying dielectric layers of semiconductor material before reaching photo conversion sites of the pixel. After they were absorbed in the semiconductor, they are converted into electronic charges, and collected in the pixel areas. After that, they are buffered to higher processing circuits.

There are many different ways to collect and convert photon energy into electronic charge. Only silicon based image sensing elements were investigated in this work. Working principles of photogate and photodiode type of photo conversion elements were explained. Pixel definitions and performance metrics of image sensors and their noise sources were reviewed. Optical limitations were also investigated for large format image sensors. Image sensor characterization and important measurement methods were explained at the last sections of this chapter.

CHAPTER 3 CHARGE-COUPLED DEVICE (CCD)

Charge-coupled device (CCD) is a silicon-based integrated circuit (IC) consisting of a dense matrix of photosensitive elements, or pixels. These pixels are either photodiodes or photogates or their variations. CCD operates by converting light into electronic charges, electrons or holes. Charges are generated by the interaction of impinging photons with semiconductor. Generated charges are collected and stored in potential well pockets formed in confined and separated one or two dimensional grids, in pixel array. Pixel charge pockets are subsequently transferred across the chip through vertical and horizontal shift registers towards common output amplifier(s) for off-chip signal processing.

CCDs were invented in late 1960's by researchers at Bell Telephone Laboratories while working on a new type of memory device. Further studies on this new device showed that it could be used as imaging element because of its ability to convert light with photoelectric effect, and transfer charge. The structure and basic applications were published in 1970, opening a new era of image sensor research [Boyle70, Amelio70]. It took eleven years though to make inexpensive, commercially available, CCD based camera systems because of several limitations [Fossum93b]. In 1980, Sony put the first 120 thousand pixels, color CCD camera on the market, [Cohen80, Melen73]. More than three decades of development has allowed the now matured CCD technology to be used in a wide variety of high performance commercial, scientific, industrial, and military applications.

3.1 CCD Pixels

Basic architecture of a CCD image sensor composes of one or two dimensional array of serial shift registers constructed by stacked, overlapped conductive layers of polysilicon separated from a semiconductor substrate by an insulating film as illustrated in Figure 3.1. CCD shift registers are used for photon conversion, charge collection, and as transport elements by applying differently biased and timed signals to the polysilicon gate electrodes. In Figure 3.1 a charge pocket is formed by applying a positive voltage to the polysilicon gate electrode (Φ2) where a p-type silicon substrate is used. Photo generated charges in and around the charge pocket are collected under the high biased CCD registers. Charges generated underneath the CCD registers that were biased low or zero volt are moved around freely in a random fashion (random walk) searching for a potential gradient to "slide" in the nearest potential well. If there is no potential gradient found nearby than the photo generated charge recombines releasing the absorbed photon energy to the semiconductor lattice.

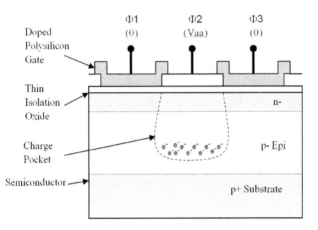

Figure 3.1. Cross section of CCD shift registers.

One of the major drawback of using CCD shift registers as an imaging element is the reflection and absorption of the impinging photons in the stacked polysilicon gates before reaching semiconductor substrate. These pre conversion losses were reduced by improving thickness and transmissivity of the polysilicon gate materials. Recent developments involve replacing the polysilicon with indium-tin oxide (ITO) that has better visible light transmissivity and provides better index matching with the silicon substrates [Meisenzahl00].

Another method for converting and collecting photo generated charges in the substrate is to use a different type of photo conversion element in the pixel and use CCD registers only for charge transport. A reverse biased photodiode is typically chosen to do this job in modern CCD image sensors. Cross section of a conceptual depiction of pinned photodiode type CCD pixel with lateral anti-blooming/reset and transfer gate is shown in Figure 3.2. CCD shift registers are used for transfer pixel charges across the CCD chip towards the output amplifier(s) after the charge pockets of photodiodes are transferred to the CCD registers. Two additional transistors are placed in each pixel. One is used to transfer pixel charge from photodiode to CCD shift register (TX). Other is used to reset the photodiode node (RST/AB). It is also used as an antiblooming barrier to reduce the overflow of access charge from one pixel to another or to the CCD shift registers. An extra layer of light shield required protecting the CCD shift registers from the impinging photons. It allows independent charge integration and transfer operation in a single CCD pixel. In modern CCD pixels, as depicted in Figure 3.2, pixel photodiode is buried under the surface with a pinning p+ implant layer improving dark current and short wavelength response. Today almost all the CCD image sensors were fabricated with pinned photodiode type pixels with more advanced transfer and anti-blooming structures.

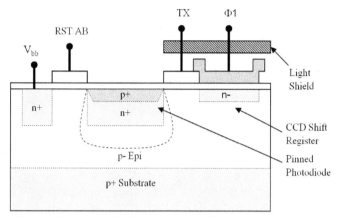

Figure 3.2. Cross section of pinned photodiode type CCD pixel and shift register.

3.2 CCD Types and Clocking Schemes

Charge is converted either under the CCD shift registers or in a separate photo conversion and collection site in each pixel before pixel readout occurs. If the photo generated charges are converted and collected in reverse biased photodiodes, than their contents are transferred to CCD shift registers first. Later, these CCD shift registers transfer the local charge pockets throughout the pixel array to one or more common charge sensing amplifier(s) that is physically separated from the pixel array. Several clocking schemes were developed for CCDs to detect and transfer charge pockets. Five of them are discussed in the following sub-sections.

3.2.1 Four-Phase CCD

Four-phase CCD, and its clocking scheme is illustrated in Figure 3.3. Four CCD shift registers form one imaging pixel. During the scene integration period (T[0]) two of the CCD shift registers are biased low ($\Phi 1, \Phi 2$), and two are biased high ($\Phi 3, \Phi 4$). Charge pockets are formed only under the shift registers that are biased high. Thus, only the photo generated charge in and around these charge pockets are collected, and stored during integration. After the integration phase ends, CCD shift registers are clocked such a way, so that (between time periods T[1] and T[4]) to move one pixel content to the next as illustrated in Figure 3.3. It takes four clock cycles to transfer one pixel charge to the next. The entire process is repeated until all the charge packets in the imaging array have reached to the common output amplifier node. Pixel fill factor of the four-phase CCD with two out of four phase biased high at any given time is 50%. This can be increased to 75% by biasing three out of four CCD phase to high during scene integration instead of two with the expense of more complicated clock generation during and after integration period.

39

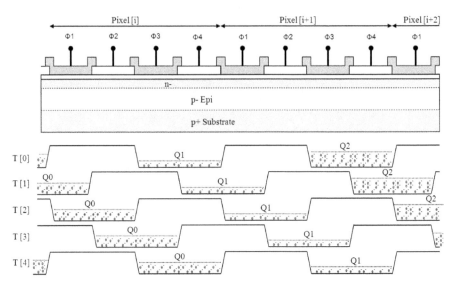

Figure 3.3. Four-phase CCD and its clocking scheme.

3.2.2   Three-Phase CCD

Three-phase CCD and clocking scheme is similar to the four-phase CCD as discussed in previous section. It is illustrated in Figure 3.4. Difference between three and four-phase clocking is that the pixel pitch is reduced in three-phase CCD structure, improving the special resolution of the CCD image sensor. In addition, only the polarity of one out of three CCD register's electrode was altered at any given time allowing high speed and low power operation. If the same CCD shift register size and doping levels used, three-phase CCD achieves lower pixel fill factor and well capacity than that of the four-phase CCD.

With the timing diagram shown in the Figure 3.4, pixel fill factor is around 33% for tree-phase CCD. Pixel fill factor can be increased by biasing two out of three of the CCD gate electrodes to high during the integration period, and generating more complicated timing diagram than that of illustrated in the Figure 3.4. In this case, the fill factor can be increased to 66%. It is still less than that of the four-phase CCD which is in best 75%. Timing and amount of pulse overlaps between the phases in three-phase CCD is more complicated than that of the four-phase CCD. Despite all these issues, three-phase CCD clocking scheme is widely used in modern CCD image sensors because of the reduced pixel size and the simplified clock drivers. It is also a fact that minimum number of gates required to implement a CCD charge transport element, in the simplest form, is three: one gate for charge storage, one for charge separation, and one for to force charge transport through the CCD.

40

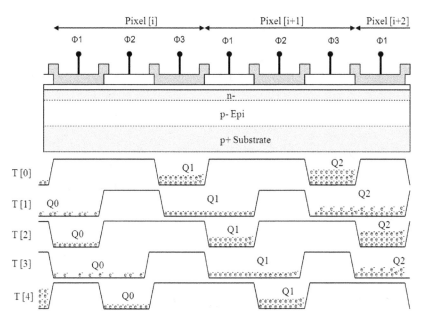

Figure 3.4. Three-phase CCD and its clocking scheme.

3.2.3 Pseudo Two-Phase CCD

Pseudo two-phase CCD and clocking scheme is illustrated in the Figure 3.5. Only two-phase clock signals are used for charge transfer. Four CCD registers form one pixel like the four-phase CCD. Each phase clock is connected to two adjacent CCD gate electrodes. Every other CCD channel underneath the polysilicon gate is doped with different level of impurity forming a build in potential step for photo generated charges. When one of the phases is biased high, and the other biased to low. Photo generated charges under half of the off and half of the on CCD registers slides into the charge pocket, effectively increasing the fill factor of the CCD pixel during scene integration. Potential step or "the electron slide" moves when the polarity of the phases altered as shown in Figure 3.5. Potential step could also be formed by having thicker gate oxide separation for half of the CCDs that are connected to the same phase. In this case, non-overlapping clock phases should be generated. Phase overlaps are very critical for oxide adjusted potential steps while it is not important for doping adjusted slides in pseudo two-phase CCD. Major drawback of pseudo two-phase CCD is the added fabrication cost and complexity to form proper potential slides under the CCD gates. In addition, total amount of capacitive load of the clock phases is doubled requiring stronger, more power hungry clock drivers to drive the clock signals. Pixel fill factor is 75% for pseudo two-phase CCD. Another issue with two-phase CCD clocking schemes is that it allows only

41

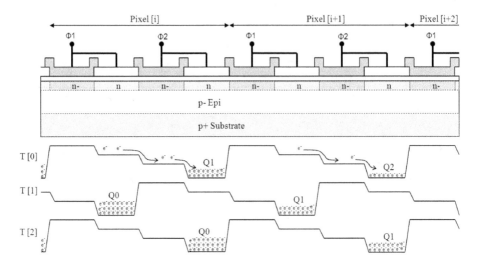

Figure 3.5. Pseudo two-phase CCD and its clocking scheme.

one directional charge transfer, left to right. In four- and three-phase CCD, charge can be transported both directions by altering the clocking phase sequences.

3.2.4 True Two-Phase CCD

True two-phase CCD and its clocking scheme are illustrated in Figure 3.6. Three step potential slides for photo generated charges are formed by implanting different doping levels of impurities underneath the CCD registers. The CCD register biased high (Φ2) would have the deepest charge well underneath the highest doping region (n) (time T[0]). While the polarities of clock phases were altered, potential slide shifts towards the right allowing charge pockets to move completely from one highly doped pixel region to another. Unlike the pseudo two-phase CCD, two separate doping formed underneath each CCD registers help reducing the pixel pitch and cuts the clock driver loads into half. Because of that, true two-phase CCD is used in low power, high speed and large format CCD image sensors. Main disadvantage of true two-phase CCD and clocking scheme is the increased fabrication complexity, and higher manufacturing cost.

3.2.5 Virtual Phase CCD

Virtual phase CCD and its clocking scheme is illustrated in Figure 3.7. Number of doped layers is increased significantly in virtual phase CCD [Hynecek80, Hynecek81]. Only one clock and polysilicon electrode is used for charge transfer between CCD registers and the pinned photodiode regions. Since there is no polysilicon layer between two CCD register, the virtual phase

42

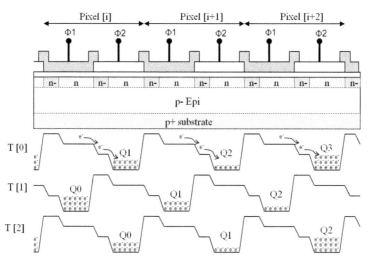

Figure 3.6. True two-phase CCD and its clocking scheme.

CCD pixels are more sensitive at both visible and blue spectrum than that of the other types of CCD. Three different levels of n-type implantation are used in virtual phase CCD along with an extra p- type implant to form the surface pinning layer. Thus, it has the most complicated and expensive fabrication process among the CCD structures. Its pixel pitch is comparable to the true two-phase CCD, yet, it has faster readout and lower power consumption among all CCD.

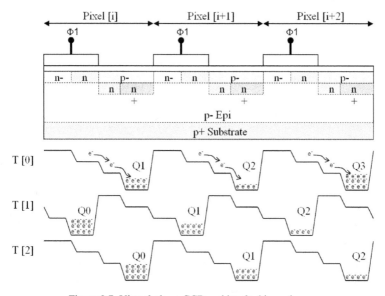

Figure 3.7. Virtual phase CCD and its clocking scheme.

43

In virtual-phase CCDs, only one clock phase is used. This feature makes these devices more reliable, faster, and cost effective (high yield). However, their fabrication processes are unique and much difficult than that of the multiphase CCDs because of the virtual-phase implant structures. In addition, relative location and global uniformity of the virtual phase implants are very crucial for high quality imaging.

3.3 CCD Readout Architectures

In a CCD image sensor, optically generated signal charges are collected under properly biased electrodes during the signal integration period. After integration period, signal charge packets are transferred from stage to stage towards readout structures. Four basic and commonly used two-dimensional readout schemes and architectures are discussed in this section. They are full frame (FF), frame-transfer (FT), interline-transfer (IT) and the frame interline-transfer (FIT) CCDs.

3.3.1 Full-Frame CCD (FF-CCD)

Full-frame transfer CCD (FF-CCD) image sensor composes of vertically tiled rows of polysilicon gate materials forming CCD shift registers as illustrated in Figure 3.8. In FF-CCD, shift registers are both used for imaging and as charge transport elements. Charge is sensed underneath the biased vertical CCD registers during integration period. Horizontal pixel boundaries are defined by the CCD registers biased low. Vertical pixel boundaries are defined by the channel stop implants. After the integration period, pixel charge pockets transported vertically one row at a time onto the horizontal shift registers. Horizontal CCD shift registers transport charges to the output amplifier. During horizontal shift operation, vertical shift registers stop, waiting entire horizontal shift register charge pockets to be read. After row read was completed, next row is shifted on the horizontal shift registers. This operation repeats until all the pixel array rows have been read.

During horizontal and vertical shift operations, pixels continue to integrate impinging photons, causing intended frame to be corrupter. This is called smear and is the major problem with the FF-CCD. Smearing of image depends on the ratio between the time required to move one frame to the CCD output amplifier(s) and the integration time. Since integration time and frame shift time are comparable, FF-CCD image sensors require a mechanical shutter to be opened during the integration time, and to be shut during the frame read time.

Because of their high fill factor, and high quantum efficiency, FF-CCDs are widely used in scientific, and studio photography, or other application in where mechanical shutter is needed. With backside illumination, a high spectral response (up to 90% quantum efficiency) and 100% fill factor with sub electron noise performance has been achieved with FF-CCD architectures. Because of their simplicity and compactness, large array versions are available with large pixel size.

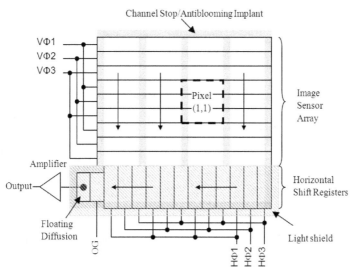

Figure 3.8. Full-frame CCD image sensor architecture.

3.3.2 Frame-Transfer CCD (FT-CCD)

Frame-transfer CCD (FT-CCD) image sensor works the same way as the FF-CCD with improved electronic shuttering capability. Main difference is that FT-CCD architecture composes of two identical CCD arrays side by side as illustrated in Figure 3.9. One CCD array is used for image sensing and transport, and the other one is used for frame storage and transport. Frame storage side of the CCD shift registers are controlled with separate clock drivers than the imaging side of the CCDs. They also covered with light shield to protect their contents. Charge is sensed during integration period in sensing area. They are transported to the frame storage area as quickly as possible at the end of frame integration. Frame storage array is read at slower speed through horizontal shift registers row by row while the image sensing array integrates next frame.

Smearing of the image depends on the time required to move the imaged frame from sensing area CCDs to the storage area CCDs. Image smear could be further reduced by adopting dual side frame transfer architectures in where half of the frame storage region is placed lower, and the other half placed at the upper part of the image sensing region. At the end of the integration period, the upper half of the frame is shifted to the upper storage region, and lower half to the lower. This architecture is called split frame transfer CCD (SFT-CCD)) architecture. In this architecture, image smear is reduced significantly because frame transfer time is reduced to half. Major drawbacks of FT-CCDs are the required extra chip area for frame storage and power to read the imaged frame. The chip area and power consumption is doubled compared with the FF-CCDs.

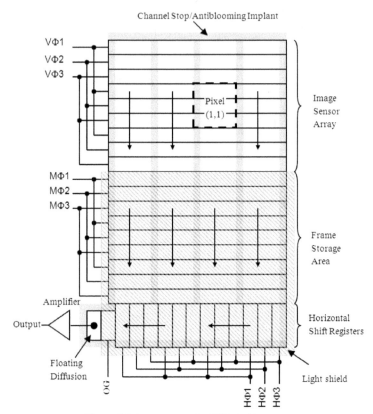

Figure 3.9. Frame-Transfer CCD architecture.

3.3.3 Interline-Transfer CCD (IT-CCD)

In interline-transfer CCD (IT-CCD) image sensor, the light blind frame storage CCD array of FT-CCD is placed between columns of pixels as illustrated in the Figure 3.10. Each pixel composes of a photodiode or photogate type photo sensing element and a set of vertical shift registers, and other switches. After integration period, pixel charge pockets are transferred locally to a set of vertical shift registers in one clock cycle. After the integrated frame is transferred to the vertical shift registers, pixel array is reset, and the next frame integration starts. During the frame integration, previous frame stored in the vertical shift registers is shifted row by row to the horizontal shift registers and to the output amplifier(s). IT-CCD sensor with pinned photodiode pixel is widely used in today's image sensor applications especially in mid to high-end consumer, industrial, scientific and military applications. IT-CCD provides best chip size and power consumption performance among the other CCD types.

46

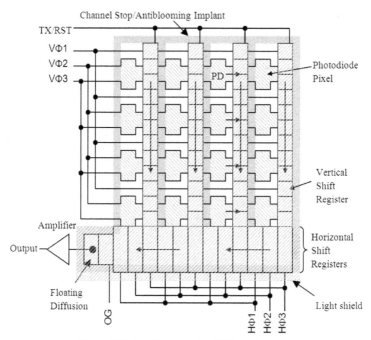

Figure 3.10. Interline-transfer CCD architecture.

3.3.4 Frame-Interline-Transfer CCD (FIT-CCD)

Frame-interline-transfer (FIT) CCD architecture is a combination of IT-CCD and FT-CCD architectures as illustrated in the Figure 3.11. FIT-CCD image sensor composes of imaging array structures of the IT-CCD, and frame storage structures of the FT-CCD. Pixel charge pockets are transferred to the vertical shift registers in the imaging array in one clock period like in the IT-CCDs. Vertical shift register contents are then shifted quickly to the frame storage area like in the FT-CCD. This operation reduces the image artifacts that caused by the slow vertical shift operation in IT-CCD sensors. Frame storage area is read through the horizontal shift registers at slower speed while pixels are integrating next frame. This type of CCD sensors are used in broadcast cameras to reduce streaks.

3.4 CCD Output Amplifier

There is an output amplifier at the end of the horizontal shift register. It converts electronic charge to voltage signal. A floating diffusion is used as output structure as depicted in Figure 3.12. Output structure consists of two MOSFET transistors, an n+ type floating diffusion (FD) node, and

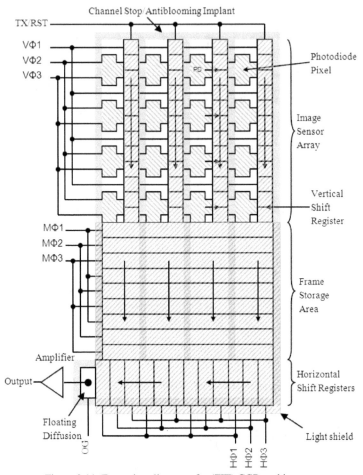

Figure 3.11. Frame-interline-transfer (FIT) CCD architecture.

a transfer gate between floating diffusion (FD) node and the horizontal shift register. One of the MOSFETs is connected between supply rail (V_{AA}) and the FD node. It is used for resetting the FD node to a known voltage before charge is transferred to FD node. Second MOSFET is used for buffering the FD node voltage to off chip signal processor. This transistor along with an external load resistor (R_L) forms a source follower amplifier. Horizontal shift register charge is converted into voltage in FD node with a ratio. This conversion ratio is called conversion gain and proportional to the FD node capacitance. It follows the basic relation of $Q = C*V$ where Q is the charge on FD node, and C is the output node capacitance. V is the voltage sensed by the on-chip output amplifier operating as source follower.

48

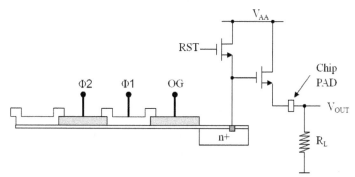

Figure 3.12. Floating diffusion type CCD output structure.

3.5 CCD Noise Reduction Methods

There are several noise sources in CCD image sensors. First noise is generated during imaging phase when the photo generated charges are collected and stored. During imaging phase not only the photo generated charges are collected and stored but also the charges generated thermally in and around of the charge collection region of the CCD pixels. Since charge generation through photoelectric and thermal processes follows the same statistics, it can be shown that the likely deviation will be plus or minus the square root of the total charge collected during imaging operation. Noise associated with the charge that is generated through photoelectric effect is called shot noise. It follows the Poisson distribution. Noise associated with thermally generated charges is called thermal noise or dark current shot noise. Another type of noise added on the pixel signal is the clocking noise generated while transferring the pixel charge pockets towards the output amplifier.

CCD pixel charges are transferred from one CCD register to the next by pulsing pixel CCD registers with non-overlapping clock phases. Total load on each clock phase are different and thus the amount of clock feedthrough to the charge pockets under each CCD registers. Clock noise arouses because of the uncertainty of the clock phase overlaps or jitter induced feedthrough on the charge pockets. It is a function of clock frequency but not the signal level, and follows the square root relation.

Other major noise sources off CCD are the electronic noises associated with the on-chip output amplifier and the camera electronics. On-chip output amplifier noise consists of two components: thermal (white) noise and flicker noise. Output amplifier thermal noise is independent of frequency and has a square root relation with the absolute temperature. Flicker noise (or 1/f noise) is strongly dependent on frequency. It originates in the existence of long-lived states in the silicon crystal, mostly the traps at the silicon-oxide interface. For a given readout configuration and

49

speed, these noise sources contribute a signal independent constant level, usually called the readout noise. The ultimate noise limit of the CCD is determined by the readout noise of the on-chip output amplifier that converts the pixel charge pockets into a change in analogue voltage.

3.5.1 Binning

Combining charge pockets of two or more CCD pixels to form a new so-called super pixel prior to readout is called "binning". Low light level imaging as a result become possible by summation of the generated charges from the single pixels. Binning improves the signal-to-noise ratio (SNR). Major drawback of this operation is that the spatial resolution of the images is reduced.

Binning is mainly used for low light level imaging and CCD-image sensors in configurations where the readout noise is dominant. Reason is that the other noise contributions do sum up during binning. Unlike read noise, dark current noise is not reduced by binning since each pixel will contribute dark current noise to the super-pixel. To ensure that dark current noise does not lower SNR during binning, it is essential that the CCD is cooled sufficiently to reduce the dark current noise to a negligible level relative to the read noise.

3.5.2 Charge Multiplication

In standard CCD image sensors, noise floor of the on-chip output amplifier determines the minimum detectable signal level. This noise is independent from signal level. It depends on readout speed. Thus, boosting the signal level before charges are transferred to the output amplifier without adding extra noise would reduce effective readout noise. Charge multiplication can be done before or after image was captured in each pixel.

Image intensifier systems, known as multi channel plates (MCP), are used for amplifying impinging photons before they reach to the CCD image sensor. This type of CCD is known as intensified charge coupled device (ICCD).

A material, known as a photocathode that emits electrons when illuminated with light is used in image intensifier system. These electrons are accelerated via electric field between the photocathode and a phosphor screen, which serves as a converter from kinetic energy stored in the electrons gained in the electric field to light energy. The phosphor screen converts electron's kinetic energy into photons through process known as cathodoluminescence. Gain of the image intensifier is the ratio between the number of excited photon from the phosphor screen and the impinging photons to the photocathode. This type of charge gain before CCD imaging array provides sub-electron readout noise.

Problems with this method are the lower quantum efficiency, worse lag and crosstalk, poor noise figure, and increased system cost. These problems were addressed in electron bombarded charge couple devices (EB-CCD), [Madan83]. In EB-CCD systems, image sensor is placed inside the vacuum tube replacing the phosphor screen with a CCD sensor allowing direct detection of the photoelectrons from the photocathode.

Another way is to use low noise charge multiplication between pixel output and on-chip output amplifier [Hynecek92] [Jerram01]. Charge multiplication is achieved by using impact ionization process that occurs when a charge travels through a very high electric field region. This high electric field regions can be formed between properly designed and biased CCD registers. The charge multiplication, and thus the charge gain per transfer, is typically in the range of one percent. This probability is increased by increasing the electric field strength via increasing the CCD phase voltage in the multiplication area. Large gain could also be achieved by increasing number of transfers and the CCD multiplier registers before the output amplifier. Such CCD imaging devices are categorized under low light level CCD (LLL-CCD) technology, and named differently by different manufacturers and researchers. Two of them are charge carrier multiplying CCD (CCM-CCD) by Texas Instrument [URL3] and electron multiplying CCD (EM-CCD) by Marconi Applied Technology (now e2V Technology) of the UK [URL4]. EM-CCD devices are available in the market with sub electron read noise performance.

3.5.3 Process Improvement

Readout noise floor of CCD can be reduced below five electrons level by improving device characteristics. Then, self-generated charge inside the sensor becomes more important, especially, the dark current. Typically CCDs are cooled to make sure that combination of the dark current and photon shot noise is smaller than the readout noise floor.

Three main sources of dark current are pronounced in CCDs. First one is the thermal generation and diffusion of the charge in the neutral bulk substrate. Second one is the thermal generation in the depletion region of the biased CCD pixel. The last and the most important one is the thermal generation due to surface states at the silicon-silicon dioxide interface below the polysilicon CCD gates. Dark current generation at this interface depends on two factors; the density of interface states and the density of free carriers that populate the interface. Originally CCDs were operated by depleting the signal channel and the interfaces resulting in high dark current generation. In modern CCD technologies, especially with MPP (Multi Pinned Phase) mode of operation, the dark current is significantly reduced by pinning the signal channel and thus the Si-SiO$_2$ interfaces pushing the dark current floor 400 times lower.

3.5.4 Design Improvement

In most cases, output amplifier noise is the dominant temporal noise in CCD image sensor. Typically, single or multistage source follower is used as output amplifier in CCD image sensors. Amplifier noise comes from active components in the signal path (thermal noise of the resistive transistor channel) and is proportional to the square root of the signal bandwidth. Thus slowing down the readout speed lowers added noise. Reducing the readout rate has its limitations too. It increases the waiting time for the pixel charge packets to reach the output stage. This effectively increases the dark current added on the pixel values and the dark current shot noise. They could be reduced by cooling the image sensor.

Another noise that is cancelled by circuit design techniques is output node KTC noise. KTC noise arises when horizontal shift register charge is sampled on the floating diffusion capacitance. This is cancelled by using correlated double sampling (CDS) technique after floating diffusion node voltage is buffered to off-chip signal processor.

KTC noise associated with the FD capacitance can be reduced by reducing the FD node capacitance by design too. In the off chip signal processor, on the other hand, sampling noise can be reduced by increasing the sampling capacitance. This is because the sampling in the FD node is done in charge domain and the sampling noise follows the square root of the KTC while in off-chip signal processor it is in voltage domain and follows the square root of the KT/C.

3.6 CCD Pixel Scaling

Very large scale CCD image sensors were introduced early 1990 for scientific application with more that 16 million pixel. Pixel size and the pixel well-capacity of commercial grade CCDs were scaled down exponentially over the past two decades as shown in the Figure 3.13. Market demand for large format image sensors is the main driving force behind this exponential shirking in pixel size over the past two decade.

3.7 CCD Spectral Response Improvement Methods

Spectral sensitivity of a CCD differs from that of a simple silicon photodiode detector because the CCD surface has polysilicon gate electrodes. These structures absorb the shorter wavelength photons and reduce the blue sensitivity of the device. A typical quantum efficiency of a consumer or scientific grade CCD is peaked at 40 percent which is lower than that of a silicon photodiode. Different methods are adapted to improve the spectral response of the CCD image sensors. Couple of them are investigated in this section.

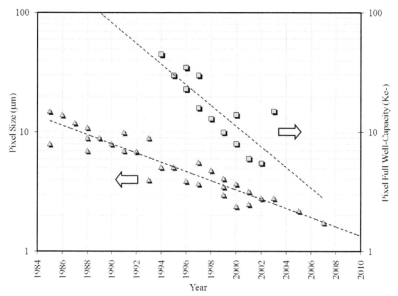

Figure 3.13. Commercial CCD image sensor scaling over two decade.

3.7.1 Frontside Illumination

Frontside illumination where incident photons must pass through above the silicon structures, especially the polysilicon gate electrodes before they can interact with the silicon substrate is used in most of the early CCD image sensors after their introduction in 1970s. However, because of the physical properties, gate materials absorb most of the blue spectrum photons that have short absorption depth. This reduces CCD's blue or ultraviolet spectrum response. The frontside illuminated CCD has virtually no response below 350 nm and less than 5% quantum efficiency at 400 nm.

CCD manufacturers have adopted variety of techniques to enhance blue and UV performance of frontside illuminated CCD. Phosphor coatings on top of gate structure, virtual phase CCD technology, open pinned phase (OPP) CCD technology [Janesick92], thin gate [Janesick94a, Janesick94b], transparent gate CCD technologies[Keenan85, Meisenzahl00, Turley92], and poly hole technique [Janesick01] are some of them.

3.7.2 Backside Illumination

Another way to improve UV response of CCD image sensor is to use backside illumination. In this technique, image sensor is illuminated from the back of the chip on where there is no gate or other attenuating structures [Shortes74]. This is one of the best ways to improve quantum efficiency of CCDs in blue spectrum. However, to be effective, the thick substrate layer of the CCD must be thinned. Thinning process is cumbersome and costly, and reduces manufacturing yield. Typically CCD substrates are thinned down to 10-20 μm.

Although backside thinning improves QE significantly, other issues appear, such as surface reflection. Air-to-silicon reflection is one of the most significant quantum efficiency losses for backside-illuminated CCDs. In UV spectrum, more than 50% of photons are reflected. This reflection could be nearly eliminated by using anti reflection (AR) coating on top of the thinned back surface. AR coating is effectively used in backside illuminated CCDs and provides almost 90% transmissivity in blue spectrum.

3.7.3 Phosphor Coating

Phosphors are wavelength converters that converts short wavelength, such as ultraviolet, into the visible spectrum in where the front illuminated silicon image sensors are more sensitive. Lumigen (a.k.a. liumogen, lumogen) is one used on frontside-illuminated CCDs [Kristianpoller64, Cowens80, Blouke80]. Lumogen Yellow S0790 from BASF, [URL1] for example, absorbs wavelengths shorter than 480nm and re-emits isotropically at about 530nm. Lumigen is deposited on top of the image sensors in elevated temperatures at sub micrometer thicknesses in vacuum. Quantum yield of the lumigen is almost 100% that for every incident photons, one visible photon is generated.

One important issue with phosphor coating is the emission direction of converted photons. Almost 50% of photons are re-emitted in the direction that they were detected by the CCD structure, while the other 50% is re-emitted to the different directions. This reduces the ultraviolet quantum efficiency to 50% of re-emitted wavelength. Lumigen is insoluble in water or most solutions, and it is readily purchased as a yellow powder in a very pure form (~99.99%). It is odorless and widely used in phosphorescent "highlight marker pens". Lumigen does not degrade optical characteristics (such as MTF, visible and NIR QE performance) of the image sensor if the coating is right over the imaging structure. This is because lumogen coating is transparent for the photons that have larger than it's characteristic wavelength. On the other hand if it is placed considerable farther away from the imaging surface, optical characteristics start degrading because of multi-directional re-emission of the visible photons. Another issue about lumogen coating is that, it evaporates under high vacuum or temperature conditions. Lumogen coating is the cheapest way to improve ultraviolet response of the CCD image sensors.

Metachrome II is another composite phosphor coating used in CCDs to improve sensitivity in blue-visible and ultraviolet (UV) wavelengths [URL2]. It dramatically improves device sensitivity in the 120-nm to 430-nm range. Metachrome II emits light at approximately 540 to 580 nm when excited with light of wavelengths shorter than 450 nm. At wavelengths longer than 460 nm, the thin layer of Metachrome II becomes transparent and, thus, has no degrading effect on the quantum efficiency of an image sensor in the visible and near-infrared portions of the spectrum. With Metachrome II coating, frontside illuminated CCDs exhibit 10% quantum efficiency below 350 nm. Backside illuminated CCDs also benefit from Metachrome II coating. Using Metachrome II coating, quantum efficiency improves by more than a factor of 10 at 250 nm.

A problem common to phosphor coatings is a steady drop in sensitivity with accumulated UV exposure. Metachrome II coated and cooled CCD cameras operating in contamination-free vacuum chambers showed no loss of sensitivity with intense UV exposure. Prolonged exposure to intense UV light did eventually produce a drop in sensitivity, which was believed to be a result of a reduction in the CCD's quantum efficiency. Typically, the stability of Metachrome II coating is guarantied for two years. On the other hand, Metachrome II coated CCD requires very low temperature for proper operation while Lumogen works in wide operation temperature including room temperature that makes it more attractive.

3.8 Charge Transfer Efficiency (CTE)

CCD sensor uniformity is generally very good. However, inefficient charge transfer may introduce shading on the image. Typical operation of a CCD requires large number of charge transfer before pixel signals reach to the readout amplifier. These transfers are accomplished by series of horizontal and vertical shifts that displace rows of charge along the chip toward a location containing the readout amplifier. If the readout amplifier is in the upper right-hand corner of a 1024 pixel by 1024 pixel CCD sensor, the charge from the pixel nearest to that corner will have to be shifted only once upward into the vertical shift register and once rightward to reach the output amplifier. On the other hand, the charge from the pixel in the lower left hand corner will have to be shifted upward 1024 times and rightward 1024 times to be read out by output amplifier. If the charge transfer efficiency (CTE) is 99.9 percent for each shift, only 12.88 percent of the charge accumulated by the lower left photodiode would remain after the required 2048 shifts, Figure 3.14. This charge loss would make the lower left corner much darker than the upper right and would also tend to blur or smear that region of the image because of spillover by charges from adjacent pixels.

Charge transfer efficiency (CTE) is improved in scientific CCD cameras by cooling and slowing the charge transfer rate. High speed charge transfer requires a different strategy. In these cameras, the read-out amplifier gain is calibrated to compensate for the charge lost by sampling extra pixels outside of the imaging area. High-performance CCD image sensors can read charge packets with only one-electron uncertainty and the charge transfer efficiency could reach 99.99995%.

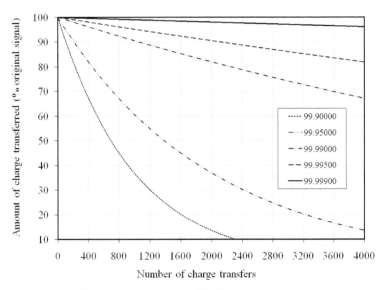

Figure 3.14. Charge transfer efficiency (CTE) of CCD image sensors.

Some control over the read-out rate as well as the size of the pixel that constitutes a sensor is permitted by scientific CCD cameras but not in the video-rate CCD cameras. Slowing the read-out usually reduces the amplifier noise associated with reading the charge. Scientific-grade CCD cameras usually offer two or more read-out rates so that speed may be traded off against noise.

3.9 Summary

CCD technology reached in its maturity, resulting in reduction in research afford over the past decade. They reached the theoretical limits in certain areas of solid-state image sensing. Amplifier noise, sensitivity, and quantum efficiency are still need to be improved for standard charge coupled device (CCD) to utilize their use at the limits.

In this chapter, question of "what makes CCDs most attractive image sensor technology?" was tried to be answered, as well as their basic operation and performance principles. Noise reduction and spectral response improvement techniques used in CCDs were investigated for the benefit of CMOS APS image sensor.

CHAPTER 4 CMOS PIXEL MODELS

Complementary metal oxide semiconductor (CMOS) pixel performance depends on fabrication process related technology parameters, and physical design rules of these processes. Best pixel performance, thus, can be achieved by optimizing fabrication process related parameters assuming best circuit design practices were utilized. However, considering most of the CMOS image sensor companies do not have their own fabrication facilities, and use foundry services, it is important to develop technology parameter based models to evaluate foundry processes and to optimize performance of pixels. In addition, these models are required for mixed-signal CMOS image sensor design methodologies to estimate pre-fabrication performance of CMOS pixel. In this chapter, few models were developed to address physical and electrical design and performance issues of CMOS pixels.

Three types of CMOS pixels were investigated for modeling. They were CMOS photodiode type passive pixel sensor (PD-PPS), photodiode type active pixel sensor (PD-APS), and photogate type active pixel sensor (PG-APS). Electrical and physical parameters of these pixels were modeled. Several CMOS processes were selected for model development and evaluation. They were the ones that have been widely used by fabless CMOS image sensor companies or research institutes over a decade. Selected CMOS processes have had minimum feature sizes between 2.0μm and 0.18μm, and supply voltages between 5.0 volt and 1.8 volt. General overviews of other type of CMOS pixels were also discussed in the later sections of this chapter.

First, a hybrid MOS threshold model equation was extracted from BSIM3v3 CMOS device model equations [Cheng98] for simple and accurate representation of back-gate bias effect on the threshold voltage [Ay05]. Then, a number of models were developed to determine electrical parameters of CMOS pixels such as pixel's reset voltage and signal range. These models were developed specifically for the photodiode type APS pixels. Pixel full-well capacity or well-depth was another modeled parameter. It was modeled based on the process technology parameters, and the physical dimensions of pixel's photosensitive element.

Second, technology design rule based physical dimension models were developed for CMOS pixels. Few full-custom pixel layouts were designed for each pixel type first. During each layout design, pixel fill factor was maximized, and pixel pitch was minimized. Best performing pixel among designs for each pixel type were than chosen for modeling their electrical parameters. Scalable critical dimensions of each selected pixel were tabulated on a table called pixel physical characteristic table (PPCT). PPCT contains generic design rule based dimensions, and offset and enlargement parameters. PPCT is used to determine pixel pitch in both horizontal and vertical directions, and total area and peripheral of the pixel's photosensitive element. Offset and enlargement parameters on the PPCT were used to change or sweep certain pixel parameters, such as pixel pitch (consequently pixel fill factor) for different process technologies.

Third, pixel property trends were determined for each pixel type under investigation. Property trends that were determined were pixel pitch, pixel fill factor, pixel full-well capacity, area and peripheral of the photosensitive region, and the parasitic loads on common output signal buses. These trends were compared with the published research and product datasheets of the CMOS pixel types to determine efficiency of designed pixel layouts. In addition, a generic pixel pitch equation for one million electron (1Me-) full-well capacity was determined for each pixel types as a function of process minimum feature size.

4.1 Hybrid Threshold Voltage Model

A CMOS active pixel sensor (APS) contains a number of NMOS transistors in each pixel. Theses transistors are used either as a switch or as an active amplifying element. In a typical three transistor (3T), photodiode type CMOS APS pixel, as shown in Figure 4.1, one transistor (M1) is used to reset the photosensitive element (PD) and other two are used as access switch (M3) and as amplifying element (M2).

A typical timing diagram of 3T CMOS APS pixel is shown in Figure 4.2. Pixel integration starts when photodiode is reset at time T1 through the reset transistor, M1. During scene integration period, photodiode node voltage (V_{PD}) drops because photogenerated electrons discharge photodiode junction capacitance (C_{PD}). During readout time period, pixel access transistor (M3) is turned on, and photodiode voltage is sampled (SHS) on a column sample-and-hold capacitance. After photodiode signal is sampled, reset transistor (M1) is turned on to start next integration period. At the same time reset level was sampled for correlated double sampling operation in column circuitry. Reset level is sampled (SHR) on column sample-and-hold capacitance at time T2. Thus, the integration time is defined by the time difference between T1 and T2.

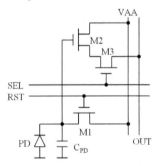

Figure 4.1. CMOS active pixel sensor (APS) schematic.

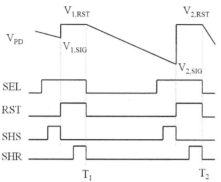

Figure 4.2. A typical 3T CMOS APS pixel timing diagram.

Reset transistor is typically sized to have the minimum allowable feature size to maximize pixel fill factor, and to reduce charge injection to the photosensitive region after reset. Also, during imaging operation, all pixel transistors have very large back-gate bias voltages which severely modulate threshold voltage of each transistor depending on the pixel's photodiode voltage. This is mainly because, NMOS transistors are used in a typical 3T CMOS APS pixel, and photodiode reset voltage is close to supply voltage, VAA. These signal dependent threshold fluctuations results in variations in photodiode reset levels and in pixel amplifier's gain.

Threshold voltage and other physical parameters of MOS transistor are well defined and modeled for circuit simulators [Shue87, Velhe93, Cheng97]. However, it become increasingly cumbersome to estimate the threshold voltage accurately by using hand calculation. It is because number of model parameters were increased dramatically when minimum channel length were shrank to deep-submicrometer levels.

A hybrid MOS threshold model equation was extracted from BSIM3v3 CMOS device model equations [Cheng98] for simple and accurate representation of back-gate bias effect on the threshold voltage, [Ay05]. In the hybrid model, it was assumed that the MOS transistor has one to three times the minimum width (W_{min}) and has a minimum channel length (L_{min}). General threshold voltage of an NMOS transistor is given with equation [4.1] [Cheng98].

$$V_{TH} = V_{TH0} + K_1 \cdot \left(\sqrt{\Phi_S + V_{SB}} - \sqrt{\Phi_S} \right) + K_2 \cdot V_{SB} + \Delta V_{TH} \qquad [4.1]$$

There are four terms in this equation. First term (V_{TH0}) is the zero back–gate bias threshold voltage, and can be determined by means of simulation or provided by foundry in a model card. It includes all of the built–in potentials and zero back–gate bias conditions for non–uniform substrate doping profile. Uncovered charges in the doping profile in the first term are covered by the second and third terms. First and second order body-effect coefficients (K_1 and K_2) are used for determining these terms. Last term, or the delta term, contains most of the short channel effects, and has significantly complicated model equations and large number of model parameter. It includes lightly doped drain (LDD) effect, narrow width with short channel expressions, charge sharing, and drain induced barrier lowering (DIBL) effects. Back-gate bias voltage (V_{SB}) is incorporated in the second, third, and forth terms. Φ_S is the surface potential, and is given for long and short channel devices with equations [4.2] and [4.3].

$$\Phi_S = \frac{2 \cdot k \cdot T}{q} \cdot \ln\left(\frac{N_{SUB}}{N_i(T)}\right) \qquad \text{Long channel devices} \qquad [4.2]$$

$$\Phi_S = \frac{2 \cdot k \cdot T}{q} \cdot \ln\left(\frac{N_{CH}}{N_i(T)}\right) \qquad \text{Short channel device} \qquad [4.3]$$

N_{CH} is channel doping concentration, N_{SUB} is substrate doping concentration, and $N_i(T)$ is temperature depended intrinsic carrier concentration of silicon. $N_i(T)$ could be determined with the equations [4.4], and [4.5].

$$N_i(T) = 1.45 \cdot 10^{10} \cdot \left(\frac{T}{300.15}\right)^{1.5} \cdot \text{EXP}\left(21.5565981 - \frac{q \cdot E_g(T)}{2 \cdot k \cdot T}\right) \qquad [4.4]$$

$$E_g(T) = 1.16 - \frac{7.02 \cdot 10^{-4} \cdot T^2}{T + 1108} \qquad [4.5]$$

It is very difficult to determine the delta term (ΔV_{TH}) in equation [4.1] with hand calculation because of the number of model parameters involved. This term is a function of parameters K_1, K_2, V_{SB}, Φ_S, W, and L. In case of fixed channel length and width, delta term can be distributed between the second and third terms of equation [4.1] with two fitting function coefficients as shown in equation [4.6].

$$V_{TH} = V_{TH0} + K_1 \cdot (1 + \eta_1) \cdot \sqrt{\Phi_S + V_{SB}} - K_1 (1 + \eta_2) \cdot \sqrt{\Phi_S} + K_2 \cdot V_{SB} \qquad [4.6]$$

The η_1 and η_2 are the fitting function coefficients (FFC). They are determined through a simple circuit simulation. They are valid for any back-gate bias voltages and limited device sizes. Simulation setup shown in Figure 4.3 can be used for determining FFCs and the device sizes that the FFCs are valid.

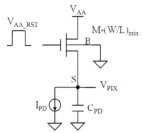

Figure 4.3. Simulation setup to determine fitting function coefficients (FFC), η_1 and η_2.

Simulation setup composes of an NMOS transistor connected between pixel supply voltage (V_{AA}) and photosensitive element (photodiode) node (node-S). Capacitor (C_{PD}) represents the diffusion and other parasitic capacitances of the photodiode. The current source (I_{PD}) represents the photo generation current. Device channel length was set to minimum allowable feature size in the selected technology while the channel width was varied with the device multiplication factor (M) during the simulation.

A transient simulation was performed to determine FFCs of a given sub-micron CMOS technology. Gate of the reset transistor pulsed from ground to V_{AA_RST} in where V_{AA_RST} is equal or larger than the pixel supply voltage, V_{AA}. After reset level is settled on node-S, reset pulse is turned off, allowing photo generation current (I_{PD}) to discharge the photodiode node capacitance (C_{PD}). During discharge, few voltage points were collected on node-S. The delta term (ΔV_{TH}) in equation [4.1] was calculated for minimum size transistor and plotted against the body factor term ($\sqrt{\Phi_S + V_{SB}}$) as shown in Figure 4.4 by using simulation data points. A first order linear fit function was then used on the data points to determine the fitting function coefficients, η_1 and η_2. The η_1 represents the slope of the linear fit line where the η_2 represents the offset.

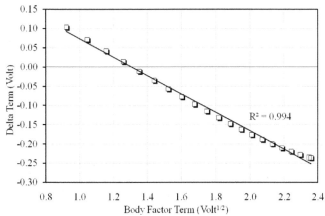

Figure 4.4. Delta term (ΔV_{TH}) versus the body factor term ($\Phi_S + V_{SB}$)$^{1/2}$.

61

After FFCs (η_1 and η_2) were determined, the threshold voltage can easily be calculated by using the equation [4.6] for any back-gate bias conditions with certain level of error. The delta term versus the device size for fixed channel length and width are shown in Figure 4.5 and Figure 4.6, respectively. The delta term less depends on the channel width than the channel length. The slope term (η_1) varies twelve percent (12%) or less for device multiplication factor of three or less (M<3) for minimum channel length NMOS transistor. Under same conditions, the offset term (η_2) varies only four percent (4%) as shown in Figure 4.7. Both terms vary almost 100% if the channel length is increased and channel width is set to minimum.

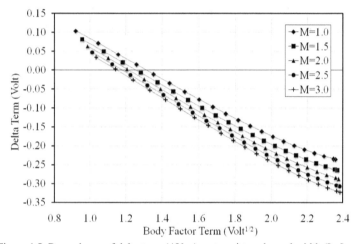

Figure 4.5. Dependency of delta term (ΔV_{TH}) on transistor channel width ($L=L_{min}$).

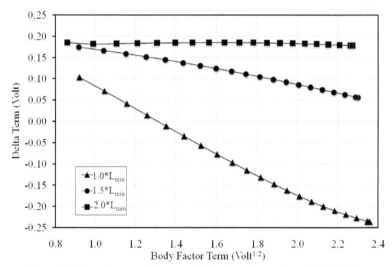

Figure 4.6. Dependency of delta term (ΔV_{TH}) on channel length ($W=W_{min}$).

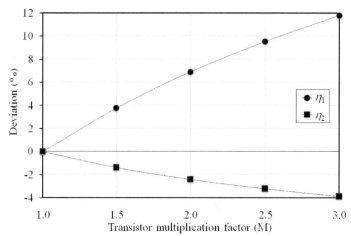

Figure 4.7. Dependency of FFCs on device multiplication factor (L=L$_{min}$)

Hybrid threshold model fitting function coefficients (FFCs) were determined for ten sub-micron CMOS processes that have been widely used by fabless CMOS image sensor companies as shown in Table 4.1. Threshold model equation [4.6] was compared with the simulation results for all these processes. Threshold model equation [4.6] resulted in better than 3% peak-to-peak calculation error in full power supply range and beyond for the back-gate bias voltages of all the CMOS processes listed in Table 4.1. Thus, equation [4.6] was used during pixel electrical model development effort described in the next section.

4.2 CMOS APS Reset Voltage Model

Photodiode reset level was modeled for CMOS APS pixel composing of a photodiode, three NMOS transistors, and a load transistor on common column bus as shown in simulation setup in Figure 4.8.

Table 4.1. Threshold fitting function coefficients (FFC) of selected CMOS processes.

	P1	P2	P3	P4	P5	P6	P7	P8	P9	P10
L$_{min}$ (μm)	0.18	0.18	0.25	0.25	0.35	0.35	0.35	0.35	0.50	0.50
V$_{DD}$ (Volt)	1.8	1.8	2.5	2.5	3.3	3.3	3.3	3.3	5.0	5.0
V$_{TH0}$ (Volt)	0.552	0.427	0.537	0.505	0.591	0.620	0.522	0.638	0.862	0.695
Φ_S (Volt)	0.859	0.880	0.840	0.860	0.859	0.898	0.865	0.847	0.838	0.865
η_1 (Volt$^{1/2}$)	-0.244	-0.054	-0.142	-0.207	-0.271	-0.112	-0.564	-0.254	-0.246	-0.062
η_2 (Volt$^{1/2}$)	-0.236	0.080	-0.122	-0.008	-0.260	-0.032	-0.555	-0.186	-0.223	0.010

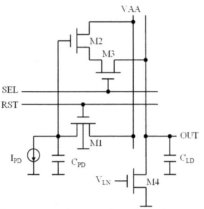

Figure 4.8. CMOS APS pixel simulation setup.

Load (M4) and buffer (M2) transistors form a source follower amplifier when pixel select transistor (M3) was turned on. Photodiode (PD) node is reset to a known pixel voltage (V_{AA}) through the reset transistor (M1). Transistor M1 turns off when pixel PD node voltage reaches one threshold below the driving supply voltage (V_{AA_RST}) of reset pulse, RST. Typically, reset transistor (M1) is designed to have a minimum device size to increase pixel fill factor of the pixel and reduce charge injection to the PD node after reset. Thus, pixel reset level could be calculated accurately by using the hybrid threshold voltage equation [4.6]. Back-gate bias voltage of reset transistor equals to the photodiode (PD) node voltage in all conditions. Threshold voltage of the reset transistor ($V_{TH,M1}$) is also modulated by the photodiode node voltage and changes when the pixel-reset signal is activated. Reset transistor turns off when the photodiode node voltage exceeds its threshold voltage minus the driving supply voltage (V_{AA_RST}) of reset pulse.

$$V_{PD,RST} = V_{SB,M1} = V_{AA\_RST} - V_{TH,M1} \qquad [4.7]$$

Photodiode reset voltage and threshold voltage of the reset transistor can be found by substituting equation [4.7] in [4.6] and solving for $V_{SB,M1}$. It can be found as:

$$V_{PD,RST} = V_{SB,M1} = \Psi^2 - \Phi_s \qquad [4.8]$$

$$\Psi = -\xi + \sqrt{\xi^2 + 2 \cdot \xi \cdot \frac{(1+\eta_2)}{(1+\eta_1)} \cdot \sqrt{\Phi_s} + \Phi_s + \frac{2 \cdot \xi \cdot (V_{AA\_RST} - V_{TH0})}{K_1 \cdot (1+\eta_1)}} \qquad [4.9]$$

$$\xi = \frac{K_1 \cdot (1+\eta_1)}{2 \cdot (1+K_2)} \qquad [4.10]$$

64

Equation [4.8] defines the reset voltage of the pixel photodiode for given technology parameters, FFCs, and the V_{AA_RST} level. Maximum achievable photodiode reset voltage is set by the pixel supply voltage (V_{AA}). This can be achieved by setting V_{AA_RST} higher than the V_{AA}.

Reset level of the photodiode and threshold voltage of reset transistor with minimum feature size were simulated by using the simulation setup shown in Figure 4.8, and calculated by using the reset model equations. First, photodiode reset level ($V_{PD,RST}$) was calculated by using reset model equation [4.8]. Than, the threshold voltage of the reset transistor was calculated by using the equation [4.7]. Finally, error was determined between simulation and calculation results for both terms. Pixel reset pulse, RST, was driven by different voltage levels (V_{AA_RST}) allowing reset transistor to charge pixel photodiode node up to pixel supply voltage (V_{AA}). Pixel reset voltage calculation error results is shown in Figure 4.9 for CMOS processes listed in Table 4.1. Error represents the difference between photodiode reset level calculated by using equation [4.8] and photodiode reset level found by using the circuit simulator. The reset model equation [4.8] estimates the pixel reset level with better than 6% (peak-to-peak) accuracy for different reset pulse levels (V_{AA_RST}) and process technologies.

Threshold voltage calculation error results are shown in Figure 4.10 for CMOS processes listed in Table 4.1. Error represents the difference between the threshold calculation by using equation [4.7] and the threshold level found by using the circuit simulator. Threshold model equation [4.7] estimates the threshold voltage with better than 2% (peak-to-peak) accuracy for different reset pulse levels, and processes.

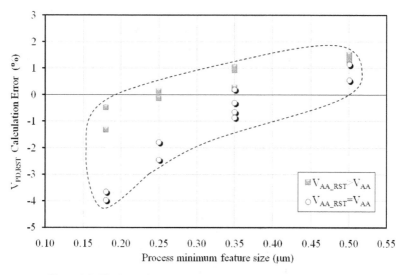

Figure 4.9. Pixel reset level calculation error versus CMOS processes.

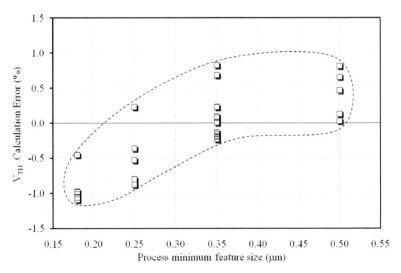

Figure 4.10. Threshold voltage calculation error versus CMOS processes.

Calculated reset voltages for the CMOS processes listed in Table 4.1 are shown in Figure 4.11. If driving supply voltage level of the pixel reset signal is set to pixel supply voltage level, ($V_{AA\_RST}=V_{AA}$) then almost 30% of the pixel signal range is lost due to the incomplete reset operation. This signal range loss could be recovered by boosting the reset pulse level (V_{AA_RST}) to higher than the pixel supply voltage (V_{AA}). Boosting amount was described as a fraction of zero-bias threshold voltage (V_{TH0}). It was called boosting factor (B). Optimum boosting factor to reset photodiode node to pixel supply voltage is shown in Figure 4.12 for the CMOS processes given in Table 4.1. Boosting factor was fount to be in between 1.5 and 2.0 times the zero-bias threshold voltages of the processes. In general, relation between minimum feature size, supply voltage and the boosting factor was found to be:

$$B = \frac{1}{4} \cdot V_{AA} \cdot L_{min} + 1.5 \qquad\qquad [4.11]$$

4.3 CMOS APS Minimum Pixel Voltage Model

In 3T photodiode type CMOS APS pixels, pixel signal range is limited by the pixel amplifier's ability to buffer photodiode node voltages. Buffering high photodiode node voltages are not a limiting factor for the source follower amplifier, if NMOS devices were used in the pixel. Highest photodiode node voltage is set by pixel supply voltage (V_{AA}) and driver supply voltage (V_{AA_RST}) of the reset pulse. Lowest input limit is set by the threshold voltage of buffer transistor (M2). It turns off when pixel photodiode voltage drops below threshold voltage of this transistor.

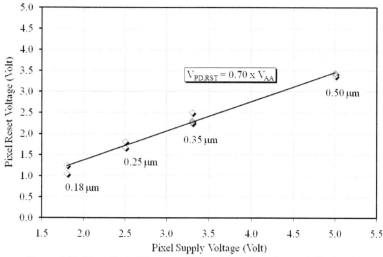

Figure 4.11. Photodiode FD reset voltage with no boosting on RST signal.

Large signal input-output transfer function of the pixel source follower amplifier could be written when the transistors M2 and M4 are working in saturation with the equation [4.12].

$$V_{OUT} = V_{PIX} - V_{TH,M2} (@ \ V_{SB,M2} = V_{OUT})$$ [4.12]

In this equation, threshold voltage of the buffer transistor, M2, is modulated by the output voltage. Considering this effect, the output voltage can be written as a function of the pixel photodiode node voltage with the following equations.

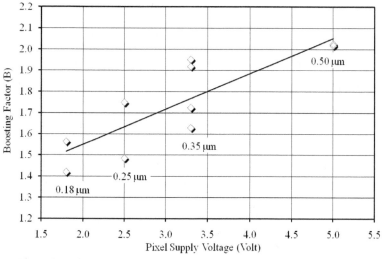

Figure 4.12. Optimum reset pulse boosting factor versus pixel supply voltage.

$$V_{OUT} = V_{SB,M2} = \Omega^2 - \Phi_S \qquad [4.13]$$

where Ω equals to the following equation.

$$\Omega = -\xi + \sqrt{\xi^2 + 2 \cdot \xi \cdot \frac{(1+\eta_2)}{(1+\eta_1)} \cdot \sqrt{\Phi_S} + \Phi_S + \frac{(V_{PIX} - V_{TH0})}{(1+K_2)}} \qquad [4.14]$$

Absolute minimum photodiode node voltage ($V_{PIX,MIN0}$) that can be buffered by the source follower can be found from the equation [4.13] by setting the output voltage to zero and extracting the pixel voltage (V_{PIX}). Linear gain range of the pixel source follower, on the other hand, ends at the operation point where the output voltage equals to overdrive voltage range of load transistor (M4). Thus, the $V_{PIX,MIN0}$ value could be calculated with the following equation.

$$V_{PIX,MIN0} = 2 \cdot \xi \cdot \left[\frac{\eta_1 - \eta_2}{1 + \eta_1} \right] \cdot \sqrt{\Phi_S} \cdot (1 + K_2) + \Phi_S + V_{TH0} \qquad [4.15]$$

The minimum pixel photodiode node voltage can be found by adding the overdrive voltage of the load transistor to equation [4.15].

$$V_{PIX,MIN} = m \cdot V_{TH0} + V_{PIX,MIN0} \qquad [4.16]$$

where m is the overdrive voltage ratio of the load transistor and is equal to;

$$m = \frac{V_{GS} - V_{TH0}}{V_{TH0}} \qquad [4.17]$$

Minimum photodiode node voltages for the processes listed in Table 4.1 were calculated by using equation [4.16] and compared with the simulation results. Simulation was performed by using the setup shown in Figure 4.8 with minimum feature size transistors. During simulation, overdrive ratio of the load transistor was set between 0.1 and 0.8, effectively setting the column bias current between 1 and 25 μA as shown in

Figure 4.13. It was found that the overdrive ratio (m) should be set to about 0.3 and 0.5 to have 5 μA and 10 μA column current, respectively.

It was found that the equation [4.16] estimates the minimum pixel voltage that can be linearly buffered by the pixel source follower amplifier with ±7% accuracy over wide range of process technologies and column bias currents as shown in

Figure 4.14 for column bias currents of 5 μA and 20 μA.

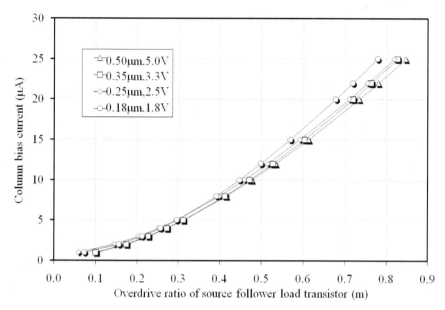

Figure 4.13. Overdrive ratio (m) of the source follower load transistor versus column bias current of process technologies listed in Table 4.1

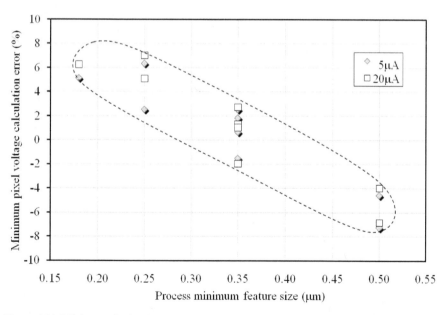

Figure 4.14. Minimum pixel voltage calculation error by equation [4.16] versus process minimum feature size and column bias currents.

4.4 CMOS APS Pixel Signal Range Model

Pixel photodiode node voltage range that can be buffered by the pixel source follower amplifier depends on device sizes of pixel source follower, overdrive voltage of the load transistor, power supply voltage used to drive the reset transistor, and transistor's process related parameters. If there was no boosting applied to the reset pulse, than the pixel signal range could be calculated by subtracting equation [4.16] from equation [4.8]. If an optimum boosting voltage (V_{AA_RST}) applied to the reset pulse in where pixel's photodiode node is reset to pixel supply voltage (V_{AA}), than the pixel signal range can be calculated by subtracting equation [4.16] from the pixel supply voltage (V_{AA}). Thus, the pixel voltage range (V_{PR}) could be written with following equation.

$$V_{PR} = V_{PD,RST} - V_{PIX,MIN} \qquad \text{(Without boosting)} \qquad [4.18]$$

$$V_{PR} = V_{AA} - V_{PIX,MIN} \qquad \text{(Optimum boosting)} \qquad [4.19]$$

Pixel voltage ranges for the processes listed in Table 4.1 were calculated by using model equations [4.18] and [4.19] for no-boosting and optimum boosting conditions at 5μA column bias current. Supply voltage of each process, and the calculation results are shown in Figure 4.15. If no boosting was applied to the reset pulse, then the pixel voltage range is around 50% of the supply voltage of the process. When the optimum boosting was applied, this range increases 80% of the supply voltage. Even with 1.8 volt supply voltage, a 1.0 volt pixel voltage range could be achievable if the boosting is adapted. For no boosting condition, pixel voltage range drops around 0.5 volt.

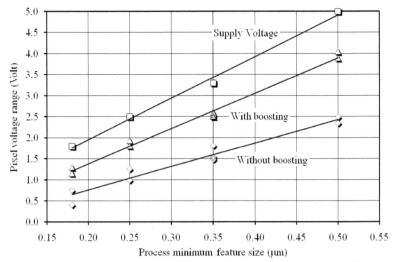

Figure 4.15. Pixel signal range for optimum boosting and no boosting conditions.

70

Pixel voltage range is related to the column current. It is because the minimum pixel voltage range is directly related to the overdrive voltage of the load transistor. Increasing overdrive voltage results in increase in the column current and reduction in the pixel signal range. Optimum bias current has to be determined for the image sensor architecture to satisfy best speed and noise performance and the pixel signal range during the design process.

Pixel signal range versus the column current for three 0.35 μm, 3.3 volt, CMOS processes listed in Table 4.1 are shown in Figure 4.16. Although these three processes were from different foundries, the pixel signal range for different column bias currents follow same trend with less than 10% variation. This variation gets smaller for smaller bias current conditions which is also preferable for low noise operation.

Reset feedthrough effect becomes significant when the pixel photodiode capacitance is comparable to the size of the reset transistor's gate to source overlap capacitance. This effect also increases when the reset signal is boosted. The feedthrough error component is given with the following equation.

$$\Delta V_{feed} = \frac{C_{OL,RST}}{C_{OL,RST} + C_{PD}} \cdot V_{AA\_RST} \qquad [4.20]$$

Larger the reset signal swing, larger the feedthrough error term becomes. This term becomes significant eventhough the reset transistor size was chosen to be the minimum feature size device for small pixel sizes. It is because photodiode capacitance shrinks with reduced pixel size. This term should be subtracted from the pixel range equations [4.18] and [4.19] in both boosted and non boosted reset cases.

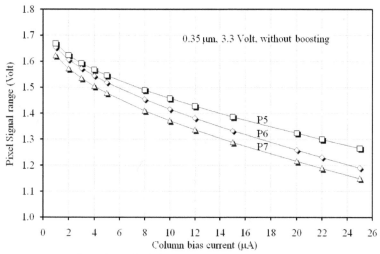

Figure 4.16. Pixel signal range versus column current.

71

4.5 CMOS APS Pixel Full-Well Capacity Model (Photodiode)

Most commonly used integration type CMOS pixels contain a reverse biased p-n junction as a photosensitive element, called photodiode. Collected charges are stored in the depletion region capacitance of the p-n junction. This capacitance is related to the area and peripheral of diffusion layer forming the p-n junction. Junction capacitances of the reverse-biased photodiode are voltage dependent. Because, depletion capacitances of p-n junction are nonlinear. Junction capacitances are function of the applied terminal voltage across the p-n junction and process parameters. Junction capacitance consists of two main components; bottom plate area capacitance and sidewall capacitances.

Bottom plate depletion region width (w_d) of a reverse-biased, abrupt p-n junction photodiode can be written as;

$$W_d = \sqrt{\frac{2 \cdot \varepsilon_{si}}{q}\left(\frac{N_A + N_D}{N_A \cdot N_D}\right) \cdot (\Phi_0 + V_{PD})} \qquad [4.21]$$

N_D and N_A are the doping concentration of n-type and p-type materials, respectively. V_{PD} is the photodiode's reverse bias voltage. ε_{si} is the permittivity of silicon, and q is the charge of an electron. Φ_0 is the junction built-in potential, and given as:

$$\Phi_0 = \Phi_T \cdot \ln\left(\frac{N_A \cdot N_D}{N_i^2}\right) \qquad [4.22]$$

where, Φ_T is the thermal voltage (26mV @ 300K), and N_i is the intrinsic carrier concentration of the material and is given in equation [4.4] for silicon. Total charge stored in the depletion region, Q_J, is given with the following equation.

$$Q_J = A \cdot q \cdot \left(\frac{N_A \cdot N_D}{N_A + N_D}\right) \cdot w_d = A \cdot \sqrt{2 \cdot \varepsilon_{si} \cdot q \cdot \left(\frac{N_A \cdot N_D}{N_A + N_D}\right) \cdot (\Phi_0 + V_{PD})} \qquad [4.23]$$

where A is the area of the p-n junction. The junction area capacitance associated with depletion region can be found by differentiating equation [4.23] with respect to the photodiode reverse bias voltage, V_{PD}. It equals to:

$$C_J(V_{PD}) = \left|\frac{dQ_J}{dV_{PD}}\right| = A \cdot \sqrt{\frac{\varepsilon_{si} \cdot q}{2}\left(\frac{N_A \cdot N_D}{N_A + N_D}\right) \cdot \frac{1}{(\Phi_0 + V_{PD})}} \qquad [4.24]$$

A more general expression for a junction with grading factor (ma), is given by

$$C_J(V_{PD}) = \frac{A \cdot C_{J0}}{\left(1 + \dfrac{V_{PD}}{\Phi_0}\right)^{ma}} \qquad [4.25]$$

where *ma* depends on the process technology. It is typically between 0.33 and 0.5. C_{J0} is the zero-bias junction capacitance per unit area. In the case of abrupt junction (*ma*=0.5), C_{J0} equals to;

$$C_{J0} = \sqrt{\frac{\varepsilon_{si} \cdot q}{2} \left(\frac{N_A \cdot N_D}{N_A + N_D} \right) \frac{1}{\Phi_0}}$$
[4.26]

Equation [4.24] shows that the value of any junction capacitance, with gradient factor (*ma*), depends on the bias voltage applied across the p-n junction, doping concentrations, area, and the temperature. This voltage will vary dynamically during pixel operation in where the photo-generated electrons are integrated in the depletion region. Typically, photodiode capacitance is defined as a large-signal, average (linear) junction capacitance, which is independent of the bias conditions. This equivalent large-signal capacitance is defined as:

$$C_{JA} = \frac{\Delta Q}{\Delta V_{PD}} = \frac{1}{V_{PD2} - V_{PD1}} \cdot \int_{V_{PD1}}^{V_{PD2}} C_J(V_{PD}) \cdot dV_{PD}$$
[4.27]

$$C_{JA} = \frac{A \cdot C_{J0} \cdot \Phi_0}{(V_{PD2} - V_{PD1}) \cdot (1 - ma)} \cdot \left[\left(1 + \frac{V_{PD2}}{\Phi_0} \right)^{1-ma} - \left(1 + \frac{V_{PD1}}{\Phi_0} \right)^{1-ma} \right]$$
[4.28]

where photodiode reverse-bias voltage across the p-n junction changes from V_{PD1} to V_{PD2}. If the pixel photodiode reset level ($V_{PD,RST}$) and the minimum linear input range of the pixel source follower ($V_{PD,MIN}$) are known than the equivalent pixel area capacitance can be written with following equation.

$$C_{JA} = A \cdot C_{J0} \cdot K_{JA} = A \cdot C_{A0}$$
[4.29]

The dimensionless coefficient K_{JA} is defined as;

$$K_{JA} = \frac{\Phi_0}{(V_{PD,RST} - V_{PD,MIN}) \cdot (1 - ma)} \cdot \left[\left(1 + \frac{V_{PD,RST}}{\Phi_0} \right)^{1-ma} - \left(1 + \frac{V_{PD,MIN}}{\Phi_0} \right)^{1-ma} \right]$$
[4.30]

The p-n junction capacitance analysis described above is applicable to sidewall junction capacitance too. Sidewall junction capacitance is formed by the n+ diffusion with doping level (N_D) and the p+ channel stop implant with doping level N_A^+. Thus, the sidewall junction-grading factor (*msw*) is different from bottom plate area grading factor (*ma*). Zero reverse-bias sidewall junction capacitance per unit area equals to:

$$C_{J0SW}^* = \sqrt{\frac{\varepsilon_{si} \cdot q}{2} \left(\frac{N_A^+ \cdot N_D}{N_A^+ + N_D} \right) \frac{1}{\Phi_{0SW}}}$$
[4.31]

where Φ_{0SW} is built-in potential for the sidewall junction. Considering the depth of the pn junction, x_J, the sidewall junction capacitance per unit length is defined as:

$$C_{JOSW} = C^*_{JOSW} \cdot x_J \qquad\qquad [4.32]$$

Total sidewall junction capacitance at zero bias can be calculated by multiplying C_{JOSW} with the perimeter of junction. Using equation [4.27], the equivalent sidewall junction capacitance, and the sidewall voltage equivalent coefficient (K_{JSW}) for voltage swing between $V_{PD,RST}$ and $V_{PD,MIN}$ can be written as:

$$C_{JSW} = P \cdot C_{JOSW} \cdot K_{JSW} = P \cdot C_{SW0} \qquad\qquad [4.33]$$

$$K_{JSW} = \frac{\Phi_{0SW}}{(V_{PD,RST} - V_{PD,MIN}) \cdot (1 - msw)} \cdot \left[\left(1 + \frac{V_{PD,RST}}{\Phi_{0sw}}\right)^{1-msw} - \left(1 + \frac{V_{PD,MIN}}{\Phi_{0sw}}\right)^{1-msw} \right] \qquad [4.34]$$

Total photodiode junction capacitance is calculated by adding equation [4.33] and equation [4.29].

$$C_{PD} = C_{JA} + C_{JSW} = A \cdot C_{A0} + P \cdot C_{SW0} \qquad\qquad [4.35]$$

Unit area (C_{A0}) and peripheral capacitances (C_{SW0}) of a number of CMOS process technologies, minimum feature sizes ranging between 2.0µm and 0.18µm, were determined using the pixel signal range models developed in the section 4.4. Results are shown in Figure 4.17. Area unit capacitance is larger for deep sub-micron devices having minimum feature size less than 0.5µm due to the increased channel stop doping level to provide better device isolation. Also effective peripheral junction area drops with reduced junction depth in deep sub-micron processes. This big difference in the capacitance results in more control on the pixel capacitance and less dark current associated with the photodiode corners, or "bird beak" of the field oxide (FOX).

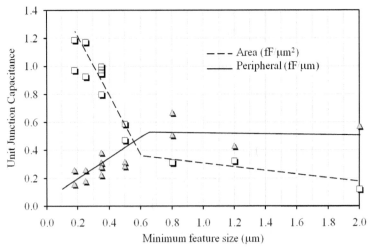

Figure 4.17. Unit junction capacitances of CMOS processes.

4.6 CMOS APS Pixel Full-Well Capacity Model (Photogate)

Photogate (PG) type APS pixel composes of a biased MOS capacitor as photo conversion element, and other switch and amplifying transistors. Pixel charge capacity is dominated by the thin gate oxide capacitance of the photogate. Other capacitances are the gate to body overlap capacitances, and the source follower's miller capacitance. Thus, the first order photogate pixel capacitance is given as:

$$C_{pix} = Area * C_{ox} \qquad (fF) \qquad\qquad [4.36]$$

Few technology parameters are important for designing a photogate type CMOS APS pixel. First one is the unit gate oxide capacitance of MOS device. Oxide capacitance versus technology minimum feature size of CMOS process technologies (minimum feature sizes ranging between 1.2μm and 0.18μm) is shown in Figure 4.18. Unit gate capacitance increases for smaller feature sizes. CMOS thin oxide capacitance with respect to the minimum feature size could be generalized with the following equation.

$$C_{ox} = 1.72 * L_{min}^{-0.843} \qquad (fF/\mu m^2) \qquad\qquad [4.37]$$

Another technology parameter is the oxide thickness that is also related to the thin oxide capacitance. Trend of CMOS oxide thickness is shown in Figure 4.19 for the CMOS processes. Relation between thin oxide thickness and the minimum feature size could be given with the following equation.

$$T_{ox} \text{ (nm)} = 1.71 * L_{min} \text{ (}\mu m\text{)} + 2.5 \text{ (nm)} \qquad\qquad [4.38]$$

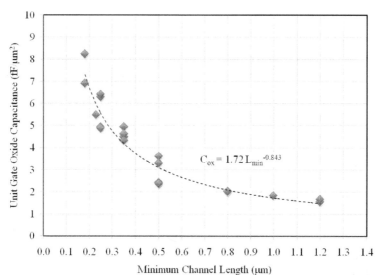

Figure 4.18. Unit gate capacitance versus minimum feature size of CMOS processes.

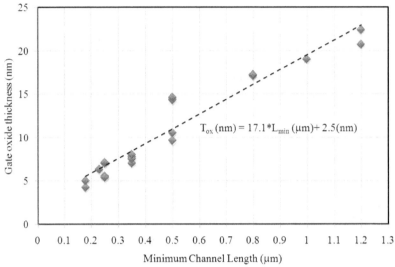

Figure 4.19. Oxide thickness versus minimum feature size of CMOS processes.

4.7 Physical Pixel Parameter Modeling

Physical design of pixel and electrical and optical properties of the manufacturing process affect overall performance of pixel. A pixel modeling methodology was developed based on CMOS foundry layout design rules, and process parameters to determine minimum achievable pixel pitch and pixel electrical properties.

First, a pixel layout was designed and optimized for minimum achievable pixel pitch on both horizontal and vertical directions by using process specific layout design rules. Then, design rule based critical layout spaces were identified that limits horizontal and vertical pixel dimensions. A number of horizontal and vertical pitch equations were determined for each pixel. Pitch equation resulting in the largest pitch value for given pixel layout and process technology was taken as the minimum achievable pixel pitch equation. Active photosensitive region was expanded on the smallest side to form a square pixel if the horizontal and vertical pitches were not the same. Expansion was done on predefined vertical and horizontal expansion lines. They are determined such a way that only the photosensitive region size is expanded, not the electrically active element sizes. Pixel photosensitive area, peripheral, and other physical and electrical property equations were extracted from optimized layouts. Using these equations and the design rules of CMOS processes, trends of pixel properties were determined for the CMOS pixel types. List of the critical layout design rules are listed in Table 4.2. These are the most commonly used layout design rules of the CMOS processes having at least three routing metal layers.

76

Table 4.2. Critical layout design rules and definitions.

Name	Layout rule definition
D0	Active to active space
D1	Active width
D2	Active overlap of active contact
D3	Contact pitch (on active)
D4	Contact pitch (on poly)
D5	Poly width (on FOX and active)
D6	Poly space (on active)
D7	Poly space (on FOX)
D8	Poly space to active (on FOX)
D9	Poly gate extension (on FOX)
D10	Poly overlap of poly contact
D11	Poly space to active contact (on active)
D12	Metal-1 width
D13	Metal-1 space
D14	Metal-1 overlap of contact
D15	Metal-1 overlap of Via-1
D16	Via-1 pitch
D17	Metal-2 width
D18	Metal-2 space
D19	Metal-2 overlap of Via-1
D20	Metal-2 overlap of Via-2
D21	Via-2 pitch
D22	Metal-3 width
D23	Metal-3 space
D24	Metal-3 overlap of Via-2

Layout design rules of few CMOS process technologies were used to determine pixel trends. Minimum channel lengths were between $0.18\mu m$ and $2.0\mu m$ for the selected processes. They were also the most commonly used CMOS processes by the researchers or by the fabless CMOS image sensor companies today and in the past. Feature sizes and the supply voltages of them are listed in Table 4.3. Process parameters were used for determining pixel properties such as pixel pitch, full-well capacity, fill factor, etc. Process parameters were extracted either from process related documents or from process model cards. These parameters were doping concentrations, junction depth, junction capacitance, threshold voltage, thickness and parasitic capacitances of the layers, optical properties, etc.

Table 4.3. CMOS process technologies used for pixel modeling.

Minimum feature size (μm)	Supply Voltage (Volt)
2.00	5.0
1.20	5.0
1.00	5.0
0.80	5.0
0.50	5.0, 3.3
0.35	5.0, 3.3
0.25	3.3, 2.5
0.18	3.3, 1.8

4.7.1 Pixel Physical Characteristic Table (PPCT) Tool

Pixel physical characteristic table (PPCT) is a tool to represent each critical dimensions of a pixel layout. It composes of number of columns and rows relating pixel layout to the generic technology specific layout design rules listed in Table 4.2. A sample PPCT is shown in Table 4.4.

Minimum horizontal and vertical pitch equations of a pixel layout are represented with $\Delta X_{min,j,k}$ and $\Delta Y_{min,j,k}$,and are given with the following equations.

$$\Delta X_{min,j,k} = \sum_{i=0}^{24} d_{i,j} \cdot D_{i,k} + x_j \cdot dx_k + y_j \cdot dy_k \qquad [4.39]$$

$$\Delta Y_{min,j,k} = \sum_{i=0}^{24} d_{i,j} \cdot D_{i,k} + x_j \cdot dx_k + y_j \cdot dy_k \qquad [4.40]$$

dx_k and dy_k are the horizontal and vertical expansion amounts of pixel layout by using the layout design rules of CMOS process k. S_k is the pixel pitch for process k. $d_{i,j}$ is the design rule multiplication factor for layout design rule number i, and equation j. One could find more than one pitch equations based on the critical distances of pixel layout. In Table 4.4, for example, only two vertical and two horizontal pitch equations were shown for a pixel layout. Rule multiplication factor for the layout was fixed for all process technologies. $D_{i,k}$ is the process design rule amount

Table 4.4. A pixel physical characteristic table (PPCT).

	Pitch				Square 1			Square2		
	ΔX_{min1}	ΔX_{min2}	ΔY_{min1}	ΔY_{min2}	W	H	P	W	H	P
S								1		2
x										
y									1	2
d0	2			1				-1		-2
d1			1	2	1				1	2
d2	4		2	1						
d3	2	2	1	2						
d4										
d5			2							
d6										
d7										
d8			4			1	2			
d9										
d10										
d11										
d12										
d13		2								
d14		4								

for rule number i, and process k. Explanation of each $D_{i,k}$ is given in Table 4.2. Design rule amount is different for different processes. x_j, y_j, and k_j are the multiplication factor for equation j. Minimum horizontal and vertical pixel pitch of given CMOS process technology could be determined by the following equations.

$$\Delta X_{min,k} = MAX \{ \Delta X_{min1,k} , \Delta X_{min2,k} , \Delta X_{min3,k} ,...\} \qquad [4.41]$$

$$\Delta Y_{min,k} = MAX \{ \Delta Y_{min1,k} , \Delta Y_{min2,k} , \Delta Y_{min3,k} ,...\} \qquad [4.42]$$

If the pixel is required to be square, than the minimum pixel pitch is defined as the maximum of the vertical and horizontal pitch, with the following equation.

$$S_{min,k} = MAX \{\Delta X_{min,k} , \Delta Y_{min,k}\} \qquad [4.43]$$

The pitch expansion amounts can be written as;

$$dx_k = EXT_k + S_{min,k} - \Delta X_{min,k} \qquad [4.44]$$

$$dy_k = EXT_k + S_{min,k} - \Delta Y_{min,k} \qquad [4.45]$$

where EXT_k is the extra expansion coefficient that can be used for pixel size sweep to determine certain parameter trends, and is defined as;

$$EXT_k = N_k * S_{min,k} + M_k \qquad [4.46]$$

where N is the pixel expansion coefficient, and M is the pixel size offset. Using these equations, the square pixel pitch for CMOS process k can be defined with the following equation.

$$S_k = dy_k + \Delta Y_{min,k} = dx_k + \Delta X_{min,k} = (1+N_k) * S_{min,k} + M_k \qquad [4.47]$$

Photo sensor area can be divided into a number of squares on the layout. In Table 4.4, for example, it was divided into two squares, marked as Square1, and Square 2. Width (W), height (H), and the effective peripheral (P) of each square can be written as a function of the design rules, pixel pitch and the expansion amounts. Total area and peripheral of the photosensitive region can be calculated by using the following equations.

$$Area_j = \sum_j W_j \cdot H_j \qquad [4.48]$$

$$Peripheral_j = \sum_j P_j \qquad [4.49]$$

4.7.2 Pixel Physical Characteristic Table (PPCT) – An Example

A two transistor (2T) CMOS passive pixel sensor layout is shown in Figure 4.20. Critical distances that limits pixel pitch on x and y directions were marked with the layout design rule numbers. Different minimum pitch equation could be obtained from the layout depending on the layout design rules and the designed physical layout. Minimum horizontal pitch for the layout can be written as;

$$\Delta X_{min1} = 2*D0+4*D2+2*D3 \tag{4.50}$$

$$\Delta X_{min2} = 2*D3+2*D13+4*D14 \tag{4.51}$$

Both equations result in a valid pitch value. However, only the largest of the two limits the minimum horizontal pitch of the pixel layout. This is true for the vertical pixel pitch equation as well. Minimum vertical pitch for the layout can be written as;

$$\Delta Y_{min1} = D1+2*D2+D3+2*D5+4*D8 \tag{4.52}$$

$$\Delta Y_{min2} = 2*D0+D1+2*D2+D3 \tag{4.53}$$

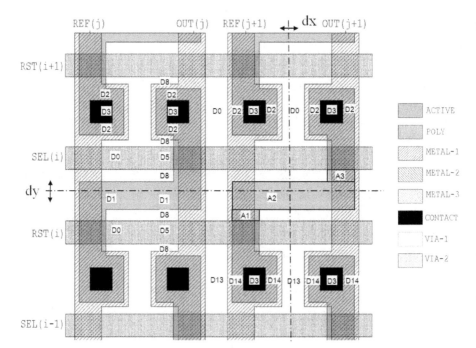

Figure 4.20. Critical dimensions of a two transistor (2T) CMOS passive pixel layout.

80

If the vertical and horizontal pitches are not equal, then the smallest of the two has to be expanded along the expansion lines (...) as shown in Figure 4.20.

Photodiode area was divided into three squares, marked as A1, A2, and A3 on the layout. Width and height of each square can be determined as a function of the layout design rules and the expansion amounts (dx and dy). Some squares would not be on the expansion lines and would have fixed width, height, or both, as in the case of area A3, and area A1. Width and height of each square could be found with the following equations.

$$A1(W)= D1 \tag{4.54}$$

$$A1(H)=D8 \tag{4.55}$$

$$A2(W)= S-D0 \tag{4.56}$$

$$A2(H)= dy+D1 \tag{4.57}$$

$$A3(W)=D1 \tag{4.58}$$

$$A3(H)=D8 \tag{4.59}$$

Peripheral of each square forming the photodiode region are given with the following equations.

$$P1=2*A1(H)=2*D8 \tag{4.60}$$

$$P2=2*[S-D0+dy+D1] \tag{4.61}$$

$$P3=2*A3(H)=2*D8 \tag{4.62}$$

All equations between [4.50] and [4.62] are represented in the PPCT for the 2T passive pixel layout (shown in Figure 4.20) on Table 4.5. This generic table is now ready to be used for different technology layout design rules.

PPCT is purely based on the layout and is different for different designs. Efficiency of the pixel layout depends on the ability of the layout designer, and the technology layout design rules. Same PPCT could be used for different processes, if the certain properties of the evaluated technologies are similar. For example, some process technologies do not allow stacked vias and contacts, and some does. Number of routing resources is another limiting factor for the pixel layout design and affects the PPCT.

Table 4.5. PPCT for 2T CMOS passive pixel layout shown in Figure 4.20.

	Pitch				Square 1			Square2			Square3		
	ΔX_{min1}	ΔX_{min2}	ΔY_{min1}	ΔY_{min2}	W	H	P	W	H	P	W	H	P
S								1		2			
x													
y									1	2			
d0	2			1				-1		-2			
d1		1	2		1			1	2	1	1		
d2	4		2	1									
d3	2	2	1	2									
d4													
d5			2										
d6													
d7													
d8			4			1	2					1	2
d9													
d10													
d11													
d12													
d13		2											
d14		4											

4.8 Passive Pixel Sensor (PPS)

Photodiode type passive pixel sensor (PPS) was first introduced in 1967 by Weckler [Weckler67, Dyck68]. A photodiode type, single transistor (1T) passive pixel sensor and a charge readout circuit is shown in Figure 4.21. Pixel consists of a reverse biased photodiode (PD), and an access transistor (M1). Pixel signal is read through a charge integration amplifier (CIA) when the access transistor is turned on. A CIA is placed at the bottom of each column. CIA keeps the column bus (COL) at a reference level (V_{REF}) and reduces the KTC noise [Noble68].

There are distributed parasitic capacitances on the column bus (C_{PAR}). These are the parasitic capacitances of the routing wires and column side diffusions of the access transistors that are not selected for readout. Each of these diffusions behaves as a small reverse biased parasitic photodiodes (PD_{PAR}) if NMOS access transistors were used. In large array sizes, they may corrupt pixel charges during readout. Column parasitic capacitances scale with the array size leading to highly capacitive column buses in passive CMOS pixels. Historically, CMOS passive pixel size scales with 10 times the minimum feature size of the process technology in use [Fossum95].

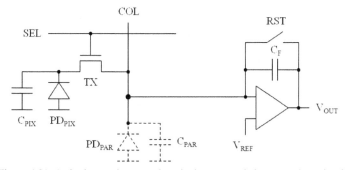

Figure 4.21. A single transistor passive pixel sensor and charge readout circuit.

4.8.1 Pixel Size Model

Single transistor, photodiode type passive pixel properties were modeled. Three different pixel layouts were designed. Two of the pixel layouts (C1 and S1) were designed assuming single metal and poly silicon layers were available for routing. Pixel select signal (SEL) was routed on the poly silicon layer while metal layer was used for column bus routing. The third pixel layout (C2) was designed assuming two metal and single poly silicon routing layers were available in the process. First metal (Metal-1) layer was used for pixel select signal routing. Second metal (Metal-2) layer was used for column bus routing. Minimum size select transistor was used in all pixel layouts.

Critical dimensions of pixel layout S1 and photodiode expansion lines are shown in Figure 4.22. Each of the critical distances that limits pixel pitch on x and y directions were marked on layout. Photodiode area was divided into three squares in layouts, marked A1, A2, and A3.

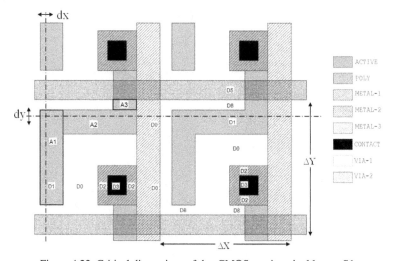

Figure 4.22. Critical dimensions of the CMOS passive pixel layout S1.

83

Table 4.6. Pixel Physical Characteristic Table (PPCT) for the pixel layout S1

	Pitch		Square 1			Square2			Square3		
	ΔX_{min1}	ΔY_{min1}	W	H	P	W	H	P	W	H	P
S											
x			1		2						
y				1	1		1	1			
d0	2	1	1	1	2	1		2			
d1	1	1		1	3		1	1	1		
d2	2	2		2	4	2		4			
d3	1	1		1	2	1		2			
d4											
d5		1									
d6											
d7											
d8		2								1	2
d9											
d10											

Pixel physical characteristic table (PPCT) for the pixel layout S1 is shown in Table 4.6. Critical dimensions of pixel layout C1 and its PPCT are shown in

Figure 4.23 and Table 4.6, respectively. Photodiode area of the layout C1 was also divided into three squares. It was found that vertical and the horizontal pitch equations of layout C1 are identical. Both layouts C1 and S1 contain one contact in each pixel. Layout C2 on the other hand contains three contacts. Contacts used in layout C2 were active to metal-1 contact, poly to metal-1 contact, and metal-1 to metal-2 via. Critical dimensions of pixel S1 and its PPCT are shown in Figure 4.24 and Table 4.7, respectively.

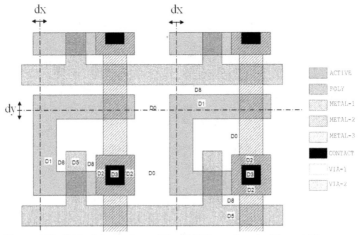

Figure 4.23. Critical dimensions of the CMOS passive pixel layout C1.

84

Table 4.7. Pixel Physical Characteristic Table (PPCT) for the pixel layout C1

	Pitch		Square 1			Square2			Square3		
	ΔX_{min1}	ΔY_{min1}	W	H	P	W	H	P	W	H	P
S											
x			1		2						
y				1	2	1					
d0	1	1	1		2						
d1	1	1	1	1	3		1		1		
d2	2	2	2	3		2		4			
d3	1	1	1	2		1		2			
d4											
d5	1	1				1		2			
d6											
d7											
d8	2	2				2		4		1	2
d9											
d10											

Minimum square pixel pitch (S_{min}) for the three layouts were determined by using the layout design rules of the CMOS process technologies listed in Table 4.3, and by using the PPCT of pixel layouts C1, C2, and S1, Figure 4.25. Layouts C1 and S1 achieve almost the same minimum pixel pitch. Layout C2 has larger pixel pitch than the others. General trend is around 7.5 times the process minimum feature size (L_{min}) minus 0.2µm for the layouts C1, and S1, and 8.0 times for the layout C2. Historically, this is around 10 times the process minimum feature size, [Fossum95].

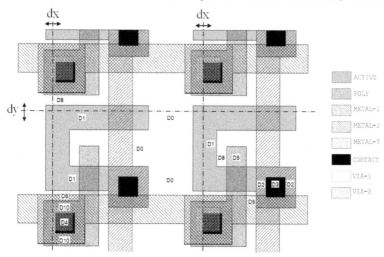

Figure 4.24. Critical dimensions of the CMOS passive pixel layout C2.

85

Table 4.8. Pixel Physical Characteristic Table (PPCT) for the pixel layout C2

	Pitch		Square 1			Square2			Square3		
	ΔX_{min1}	ΔY_{min1}	W	H	P	W	H	P	W	H	P
S											
x			1		2						
y				1	2		1				
d0	1	1		1	2						
d1	1	2	1	2	6		1		1		
d2	2					2	4				
d3	1					1	2				
d4		1									
d5	1					1	2				
d6											
d7											
d8	2	2				2	4		1	2	
d9											
d10		2									

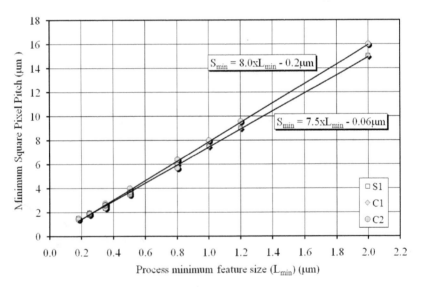

Figure 4.25. Minimum square pixel pitch (S_{min}) versus process minimum feature size (L_{min}) for passive pixel layouts C1, C2, and S1.

4.8.2   Pixel Fill Factor Model

Fill factor of pixel layout C1 with 1.0, 1.5, 2.0 and 10 times the minimum square pixel pitch (S_{min}) is shown in Figure 4.26 for different process technologies listed in Table 4.3. As shown in Figure 4.26, fill factor of the passive pixel layout C1 designed in different CMOS processes were almost the same. Same expansion was done on the other pixel layouts and it was determined that it is true for other layouts too. Thus, the pixel fill factor was determined by averaging fill factor values of different process technologies for a given pixel pitch. Resulting pixel fill factor is independent from the minimum feature size or the channel length is some cases of the process under investigation. Pixel fill factor only depends on the minimum square pixel pitch (S_{min}). Although minimum pixel pitches are similar for the three layouts, fill factor of pixel layout S1 is higher than that of the pixel layouts C1, and C2 for $S_{min}<5$.

Fill factor of the three pixels were re-calculated by expanding the pixel pitches proportionally on x and y directions marked as dx and dy on in Figure 4.22, Figure 4.23, and Figure 4.24. New pixel pitches were normalized with minimum square pixel pitch (S_{min}) of each pixel layout. Fill factor versus the square pixel pitch was plotted. Results are shown in Figure 4.27.

Pixel layouts C1 and S1 have higher fill factor than that of the layout C2. Fill factors reach 90% for pixel pitch that is five times the minimum square pixel pitch.

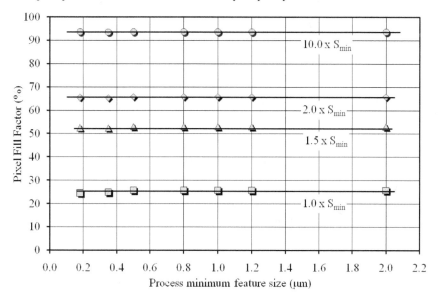

Figure 4.26. Pixel fill factor of the layout C1 with minimum pixel pitch versus the process minimum feature size.

87

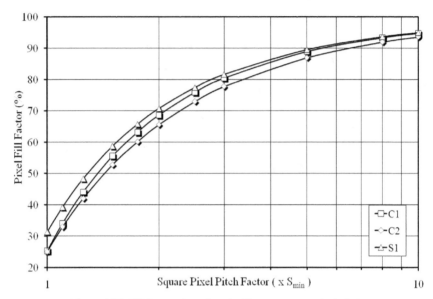

Figure 4.27. Fill factor of passive pixel layouts versus pixel pitch.

4.8.3 Pixel Full-Well Capacity Model

Pixel photodiode capacitance and hence the full-well capacity is related to the effective area and peripheral of the photodiode region, and pixel signal range. Normalized pixel pitch versus pixel photodiode capacitance of layout S1 for CMOS processes minimum feature sizes between 0.18 μm and 2.0 μm are shown in Figure 4.28. Photodiode capacitance can be determined from this graph for given technology node and S_{min} value.

For example, let's design a CMOS passive pixel, in 0.35μm 1P2M CMOS process, with 1 volt pixel signal swing and 100Ke- full-well capacity. Using equations [2.17] and [2.18] we find conversion gain to be 10μV/e- and photodiode capacitance of 16fF. Using Figure 4.28, square pixel pitch factor (S_{min}) can be found to be 2.5 for 16fF which represents fill factor of 78%. S_{min} for 0.35μm, 1P2M CMOS process can be determined by using Figure 4.25 which is 3μm. Thus, the CMOS passive pixel pitch using layout C2 style shown in

Figure 4.24 that accommodates 100Ke- full well-capacity is 2.5 times the S_{min} which is about 7.5μm x 7.5μm. This is the optimum pixel pitch that accommodates the design specification. However, if the fill factor needs to be improved, square pixel pitch factor (S_{min}) has to be increased using channel length independent graph in Figure 4.27. For example for 90% fill factor S_{min} has to be chosen to be at least 5. This makes pixel size to be 15μm x 15μm. To keep conversion gain constant pixel has to be expanded on x and y

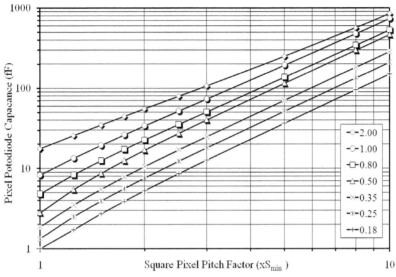

Figure 4.28. Pixel capacity versus pixel pitch of layout S1.

directions without increasing the photodiode area. Passive pixel pitch for one million electrons (1Me-) full-well capacity was determined and plotted against minimum feature size in Figure 4.29. General equation for 1Me- pixel pitch versus minimum feature size is given in equation [4.63]. In this case, it was assumed that the column charge amplifier reference voltage (V_{REF}) was set to 70% of the supply voltage (V_{AA}).

$$S_{1Me-} (\mu m)= 24 * L_{min}(\mu m) + 9\ \mu m \quad (\text{for } V_{REF} =0.7* V_{AA}) \qquad [4.63]$$

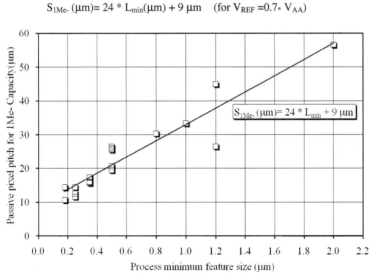

Figure 4.29. Passive pixel pitch versus minimum feature size for 1Million electron full-well.

4.8.4 Pixel Parasitic Load Model

In CMOS passive pixels, column parasitic capacitances become the main power and speed bottleneck when pixel array size is increased. These capacitances are largely related the column reference voltage, diffusion area and peripheral of the disconnected access transistors that are connected to the common column bus. Column parasitic load versus pixel array size could be calculated assuming column routing wires have lower parasitic capacitances than that of the diffusions, and area and peripheral of the off-switch diffusions were minimized. Total column parasitic capacitance for different array sizes versus the minimum process feature sizes are shown in Figure 4.30. In this plot, column reference voltage was set to 70% of supply voltage of the process technology under consideration. Using the design rules of the technologies, minimum off-switch diffusion area were determined and used.

For example, designing a 4Kx4K CMOS passive pixel array imager in 0.25μm CMOS process, designer has to deal with column bus capacitance from off-switch diffusions which is more than 2.5pF not including the wiring capacitances. This load reduces for smaller technology nodes, and increases for larger ones. Thus, it might be feasible to use CMOS passive pixels in smaller feature size technologies in large formats, considering only the loading from the off-switch capacitances on the column bus.

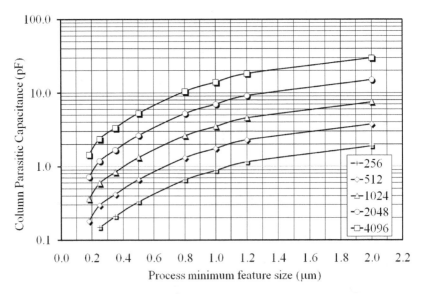

Figure 4.30. Total column parasitic capacitance from off-switch diffusions versus process minimum feature sizes for different array formats.

4.9 Photodiode Type Active Pixel Sensor (PD APS)

During the invention of the passive pixel, it was recognized that insertion of a buffer/amplifier into each pixel could potentially improve the performance of the passive pixel. A sensor with an active amplifier within each pixel is referred as an active pixel sensor (APS). Many types of CMOS and non-CMOS active pixel sensors were investigated and developed over the past decade.

Three transistors (3T), photodiode-type CMOS APS pixel hasn't been changed since first introduced in 1968 [Noble68]. It composes of a reverse biased photodiode (PD), a reset transistor (M1), and two active buffer transistors (M2 and M3) as illustrated in Figure 4.31.

Before scene integration, reverse biased photodiode is reset to a known voltage (VAA) through the reset transistor (M1). During integration period impinging photons are converted into electron-hole pairs. Electrons are stored in the reverse biased p-n junction depletion capacitance (C_{PD}) and the holes are dumped to substrate (or ground) or they are recombined. Collected electrons discharge the pixel capacitance from the pixel-reset level, depending on the strength of the illumination condition on top of the pixel. Typically, common drain (source follower) type active amplifier is used to buffer the pixel signals in CMOS APS pixels. Two pixel transistors form the upper half of the source follower while the lower part of the amplifier, or the current sink transistor located at the bottom of the column readout bus (not shown in Figure 4.31). An extra switch can be placed before the current sink transistor to save power during idle period. Standard timing diagram of a three transistor (3T), photodiode-type CMOS APS pixel is given in the Figure 4.2.

Common column readout bus has parasitic elements like the passive pixel sensor. These increase loading of the pixel source follower and slow down the readout speed. Potential-well diagram of 3T photodiode-type CMOS APS pixel is shown in Figure 4.32.

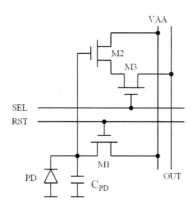

Figure 4.31. Photodiode type Active Pixel Sensor (APS) circuit diagram.

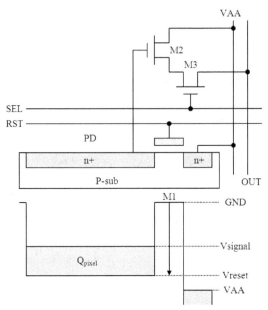

Figure 4.32. Potential-well diagram of the 3T PD-type APS pixel.

During pixel reset, M1 is turned on by pulsing reset line (RST) from ground to reset driving supply voltage (V_{AA_RST}). Photodiode (PD) node is reset one threshold voltage below this voltage depending on the length of the reset pulse.

$$V_{PD\_RST} = V_{AA\_RST} - V_{TH,M1}(V_{PD\_RST}) \qquad [4.64]$$

where $V_{TH,M1}(V_{PD\_RST})$ is the threshold voltage of the reset transistor, M1. This threshold voltage depends also on the photodiode voltage. Usually, M1 is design to have minimum device size.

Integration period starts as soon as the reset transistor turns off.)After the integration period, pixel is selected by select signal (SEL). Select transistor (M3) connects active source follower transistor (M2) to the other half of the source follower amplifier located at the bottom of the column readout bus. Photodiode signal voltage is read first. This is non-destructive voltage readout because the high impedance input node of the source follower is directly connected to the photodiode node. Typically, source follower has gain of less than unity due to the body effect of the active amplifying transistor (M2). This signal level is sampled on one of the column sample and hold capacitors in column circuitry. After photodiode signal is read, pixel photodiode is reset and sampled on another sample and hold capacitor in column circuits. Then, integration starts again for the next scene integration. Sampled pixel signal and reset levels are processed and transferred off-chip in digital or analog format during the scene integration time.

4.9.1 Pixel Size Model

A number of CMOS APS)layouts were designed to evaluate the pixel parameters. It was assumed that, for each layout, a limited number of signal routing layers were available in the process. Layouts were optimized to provide largest fill factor and minimum pixel pitch. After layouts were designed, critical dimensions that set the horizontal and vertical pixel pitches were determined with respect to the design rule distances. Vertical and horizontal expansion lines were identified for each pixel. Pixels were expanded on these lines on x and y directions. Pixel properties were then re-calculated to map the process and size related trends of the APS pixels. Pixel photodiode region was divided into sub squares. Critical dimensions of each of these sub-squares were tabulated on pixel physical characteristic table (PPCT). Area and peripheral of each sub-square were then calculated. After evaluating performance of each pixel, only two best layouts were reported in this section.

Layout as well as the vertical and horizontal expansion lines of the first APS pixel (APS2M1) is shown in Figure 4.33. It was assumed that two metal layers were available for signal routing in the process. First metal layer was used for routing supply and output signals on vertical direction. Second metal layer was used for routing pixel select and reset signals on horizontal direction. Pixel photodiode area was divided into four squares. Three of them were the main defining squares for minimum pixel pitch. The forth square appears when the pixel pitch was expanded certain amount on x or y direction.

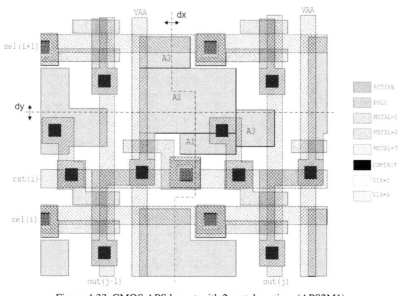

Figure 4.33. CMOS APS layout with 2 metal routing. (APS2M1)

PPCT for sub-micron technology design rules in where stacked vias, and contacts are allowed is shown in Table 4.9 for pixel layout APS2M1. Another characteristic table was also extracted and used for long channel technology design rules in where stacked via, and contacts were not allowed. At least two critical vertical and horizontal pitch values ($\Delta X1$, $\Delta X2$, $\Delta Y1$, $\Delta Y2$) were found to determine the minimum pixel pitches (ΔX, ΔY). Minimum horizontal pitch (ΔX) was determined by taking the largest of the critical horizontal pixel pitch values ($\Delta X1$, $\Delta X2$, $\Delta X3$,…). Largest of all were chosen to be the minimum pixel pitch to form a square pixel. It was achieved by expanding the pixel pitch in x or y direction on expansion lines.

The second pixel layout (APS3M1) is shown in Figure 4.34. It was assumed that process technology has three metal layers for routing. Pixel reset signal was routed on the poly silicon layer. Pixel output and supply were routed on the third and second metal layers, respectively. First metal was used for routing pixel select signal. Pixel photodiode area divided into three major squares. Two of them were the main minimum pitch defining squares.

Table 4.9. Pixel Physical Characteristic Table (PPCT) for pixel layout APS2M1.

	Pitch				Square 1			Square 2			Square 3			Square 4*		
	ΔX_{min1}	ΔX_{min2}	ΔY_{min1}	ΔY_{min2}	W	H	P	W	H	P	W	H	P	W	H	P
S					1			1	1	4	1				1	
x											1					
y																
d0				1	-2			-2		-4	-1	-1	-2	1	-2	2
d1					-1									1	-1	2
d2																
d3	1	1	1	2	-1	1	2	-1	-1	-4	-1			-1		
d4	1	2			-2	-4					-1					
d5			2		-1	3	6		-3	-6	-1	2	4	-2		
d6			1								-1			-1		
d7		1							-1	-2		1	2			
d8				1	-2	2	4		-4	-8	-2	2	4	-2		
d9																
d10	1	2				-4	-8				-2					
d11																
d12	1	1														
d13	4	3	2													
d14	3	4	2	4	-2	2	4	-2	-2	-8	-2			-2		
d15	2		4													
d16	1		2													
d17																
d18			1													

* : Calculate IF $\Delta X_{min}-[2*d0+d1+d3+2*d5+d6+2*d8+d14]>d1$

94

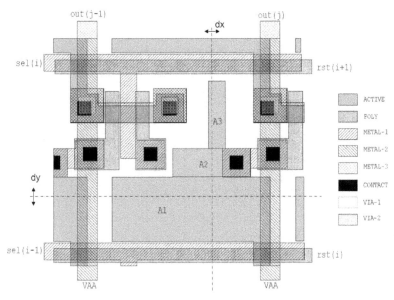

Figure 4.34. CMOS APS pixel layout (APS3M1) for process evaluation.

The third one appears if the pixel pitch was expanded certain amount in x direction. PPCT of the layout APS3M1 is shown in Table 4.10. Two horizontal and two vertical critical dimension equations were determined for the pixel layout. Largest of four was taken to be the minimum pixel pitch and pixel was expanded on x, or y-direction to form a square pixel. Photodiode active area dimensions were determined and tabulated in Table 4.10 in terms of pixel pitch, extra x and y expansion amounts (dx and dy), and the critical design rules. Some active squares appear when the pixel size expanded to certain level, like the square number three in Table 4.10. In minimum size square pixel, area three (A3) does not exist.

Pixel properties such as pixel capacitance, and fill factor were calculated by using pixel voltage range model, and pixel capacitance models developed in previous sections. Using these models, pixel fill factor and minimum pixel size versus technology minimum feature size were determined. Pixel pitch of pixel layouts APS3M1 and APS2M1 versus technology minimum feature size is illustrated in Figure 4.35. Pixel layout APS2M1 has less efficient fill factor and pixel properties than that of the pixel layout APS3M1. As seen in Figure 4.35, APS2M scales with 13.7 times the minimum feature size while the APS3M1 scales with 13.5. Pixel pitches of the published research and commercial CMOS APS image sensors versus the minimum feature size of their fabrication processes between 1997 and 2004, [REFL1] are illustrated in Figure 4.36. Historically, these three transistors (3T) CMOS APS pixels scale with 13.7 times the minimum feature sizes like the pixel layout APS3M1 with less than 0.9μm pitch offset.

Table 4.10. Pixel Physical Characteristic Table (PPCT) for pixel layout APS3M1.

	Pitch				Square 1			Square 2			Square 3*		
	ΔX_{min1}	ΔX_{min2}	ΔY_{min1}	ΔY_{min2}	W	H	P	W	H	P	W	H	P
S					1	2					1	2	
x						-1	1	1			1		
y					1	2					-1	-2	
d0	1				-1	-2					-1		
d1		1	2		1	1					-1	-2	
d2													
d3	0.5	1				-1		1	1	3	-1	-2	
d4		2	1	1									
d5			1								-1	-2	
d6			1										
d7													
d8		1	1.2	2				1	-1	-1	-2	-4	
d9			1										
d10		1	1	1				1	-1	-1			
d11													
d12	1	1	1										
d13	2	2	2	1				-1	1	1			
d14	1	5	1	1		-3		3	2	7	-2	-4	
d15	3		2										
d16	1.5		1										

* : Calculate IF dx>[d0+d1]

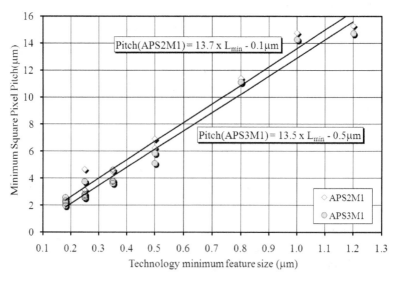

Pitch(APS2M1) = 13.7 x L_{min} - 0.1μm

Pitch(APS3M1) = 13.5 x L_{min} - 0.5μm

◇ APS2M1
◎ APS3M1

Minimum Square Pixel Pitch(μm)

Technology minimum feature size (μm)

Figure 4.35. Pixel pitch scaling of the APS pixel layouts APS2M, and APS3M1.

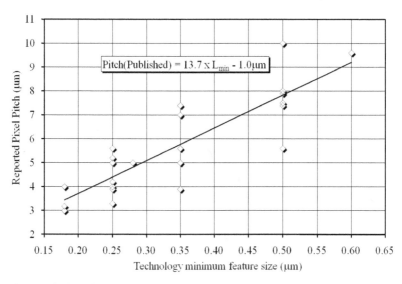

Figure 4.36. APS pixel pitch versus the process minimum feature sizes of published CMOS image sensors between 1997 and 2004, [REFL1].

4.9.2 Pixel Fill Factor Model

Fill factor of the CMOS APS pixel layouts APS2M1 and APS3M1, and the published CMOS APS pixels are shown in Figure 4.37, [REFL1]. In average, published APS pixels achieve between 30% and 35% fill factor, almost same as the pixel layout APS2M1. Fill factor of APS3M1 is between 15% and 25% because of the extra metal layer used If the pixel pitches were equalized by adding the offset differences on the pitches, both designs would achieve almost the same fill factor between 30% and 35%.

4.9.3 Pixel Full-Well Capacity Model

Pixel capacitance of the APS pixel layouts, APS2M1 and APS3M1, that result in minimum pixel pitch were determined. Their and the published pixel's capacitances [REFL1] are shown in Figure 4.38. As shown in Figure 4.37 and Figure 4.38, pixel APS2M1 is more closely emulates the published pixel fill factor and the pixel capacity.

Figure 4.39 shows the pixel well capacity versus the normalized pixel pitch for different CMOS processes that have feature sizes between 0.18 μm and 1.2 μm. Full-well capacity that holds one million electron (1Me-) is around 160.2fF. Figure 4.40 shows the pixel pitches of APS2M1 that can hold 1Me-. Pixel pitch and the minimum feature size has the following relation.

$$\text{Pitch (1Me}^-\text{)} = 13.8 \, L_{min} + 12 \, \mu m \qquad [4.65]$$

97

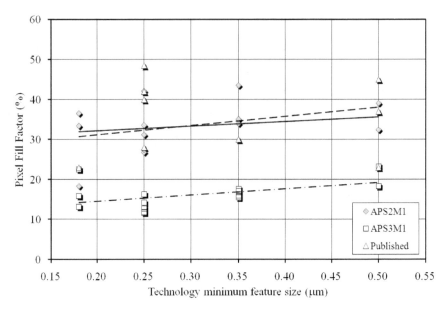

Figure 4.37. APS pixel fill factor versus the process minimum feature sizes of published CMOS image sensors between 1997 and 2004, [REFL1], and layouts APS2M1, and APS3M1.

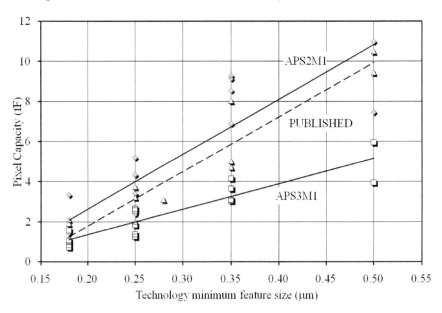

Figure 4.38. APS pixel capacity versus the process minimum feature sizes of published CMOS image sensors between 1997 and 2004, [REFL1], and layouts APS2M1, and APS3M1.

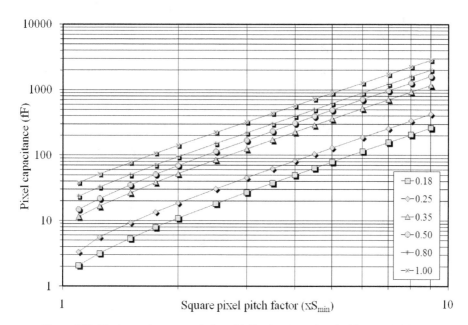

Figure 4.39. Pixel capacity versus pitch multiplication factor for pixel layout APS2M1.

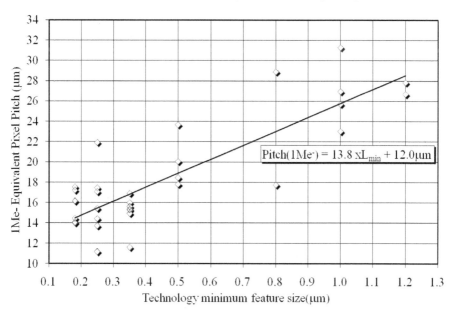

$$Pitch(1Me^-) = 13.8 \; xL_{min} + 12.0\mu m$$

Figure 4.40. Pixel pitch that achieves capacitance of 1Me- versus technology minimum feature size for CMOS APS pixel APS2M1.

4.10   Photogate Type Active Pixel Sensor (PG APS)

Photogate type CMOS active pixel sensor (APS) was introduced in 1993 [Mendis93a] for high performance and low light level scientific imaging applications, [Mendis94a, Mendis97]. Photogate APS pixel composes of five transistors as shown in Figure 4.41. Photogate transistor (M5) is used as charge sensing and integration element. Charge transfer transistor (M4) connects photogate to the sense node. Reset transistor (M1) is used for resetting floating diffusion capacitance (C_{FD}) of the sense node. Source follower active transistor (M2) and a select transistor (M3) are used for buffering the sense node voltages. Photo generated charges are accumulated under a photogate transistor (M5) during integration time. At the end of the integration time, sense node is reset first. Reset voltage is buffered by the source follower to a column sample and hold circuitry. Then, accumulated pixel charge is transferred to the sense node by pulsing the photogate. New voltage is then buffered by the pixel source follower and sampled for further processing by the column sample-and-hold circuitry.

Potential well diagram and the cross section of a photogate type CMOS APS pixel are shown in Figure 4.42. Gate of the photogate transistor is biased to form a channel depletion region to convert and collect photo generated charges. During readout period, photogate (PG) charges are transferred to a floating diffusion (FD) sense node capacitance through the transfer transistor (M4). Unlike modern CCD pixels, photogate poly does not overlap with the transfer transistor's gate poly. There is an n+ bridging diffusion region in between photogate and transfer transistor channel. Although, bridging n+ diffusion cause image lag in PG type APS pixels, it has minimal effect on circuit performance and permits use of single poly CMOS processes [Mendis94b].

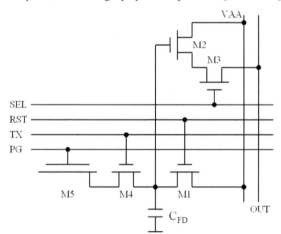

Figure 4.41. Photogate type APS pixel schematic.

100

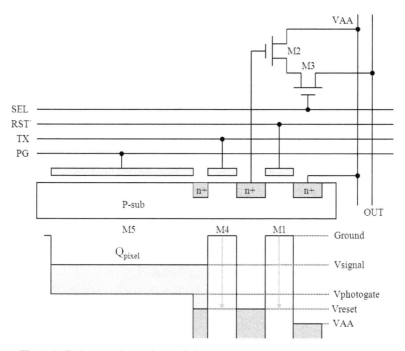

Figure 4.42. Cross section and potential well diagram of the photogate APS pixel

Typically, an n+ active region is used to form floating diffusion sense node capacitance. CMOS photogate type APS pixels suffer the early CCD pixel shortcomings, because of the poly silicon gate properties. The photogate-type APS uses five transistors per pixel and historically has a pitch equal to twenty times (20x) the technology minimum feature size [Nixon96].

4.10.1 Pixel Size Model

Photogate type APS pixel properties were modeled based on the optimized pixel layout (PG3M1) shown in Figure 4.43. In this design, sense node capacitance was maximized, but could be adjusted depend on the design specifications. During layout design, photogate area, fill factor, and sense node diffusion area were maximized, and pixel pitch was minimized. All transistors have had minimum device sizes. It was assumed that the process has three metal and single polysilicon layers for routing. Also, it was assumed that stacked vias and contacts were allowed. Four horizontal signals were routed on metal-1, metal-2 and poly layers. The metal-3 layer was used for routing pixel power and output lines on vertical direction. Like the photodiode type APS pixel, number of process technology design rules and parameters were used to determine the pixel properties. PPCT of the pixel layout is shown in the Table 4.11.

101

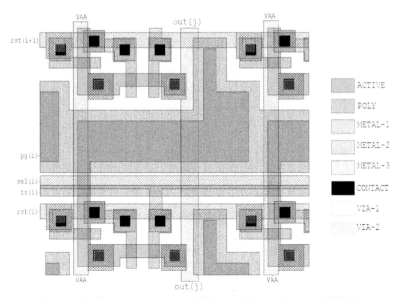

Figure 4.43. Optimized photogate (PG) type APS pixel layout (PG3M1).

Table 4.11. Pixel Physical Characteristic Table (PPCT) for pixel layout PG3M1.

	Pitch		Square 1		Square 2*		Square 3**	
	ΔX_{min1}	ΔY_{min1}	W	H	W	H	W	H
S			1	1	1		1	
x								
y								
d0			-1		-1		-2	
d1	1	2		-1	-1		-1	1
d2								
d3					-1	1	-1	
d4	2	1		-1	-2		-1	
d5		1		-1	-1	1	-1	
d6				-1				
d7	2	3		-2	-3		-2	
d8	1				-2	2	-1	1
d9	1	3		-3	-2		-1	1
d10	3	1		-1	-4		-2	
d11								
d12								
d13	1	1		-1				
d14	1	1		-1	-2	2	-2	
d15	2							
d16	1							

* calculate if [d4+d13+d10+2*d15+d16-d3-d5-d14]>[2*d0+d1]

** calculate if [d4+d13+d10+2*d15+d16-d3-d5-d14]>[d0+d1+d4+d7+d8+d9+2*d10]

102

4.10.2 Pixel Fill Factor Model

First, literature was scanned and general trend of published photogate type APS pixel pitches and their pixel-fill factors were determined to adjust fill factor of designed photogate pixel layout (PG3M1). Pixel fill factor and pixel pitch versus the technology minimum feature size of published pixel data between 1990 and 2004 [REFL2] is shown in Figure 4.44. Published pixel pitch has a trend of 18.76 times the minimum feature size of the process technologies with an offset of 1.76µm, and the fill factor is around 31% for the CMOS photogate pixels. Fill factor is one of tha main factor to deteremine the pixel pitch. Smaller pixels can be designed but with the expens of the fill factor.

Absolute minimum pixel pitch and associated pixel fill factor that can be achieved with designed pixel layout, PG3M1, is shown in Figure 4.45. Pixel pitch of this design has a trend of 13 times the process minimum feature size with an offset of 0.4µm. Pixel fill factor of the absolute minimum pixel pitch of PG3M1 is between 6% and 10%. When the pixel pitch is expanded so that pixel fill factor is more that 30%, the designed pixel pitch trends with 17 times the technology minimum feature size with an offset of 0.5µm as shown in Figure 4.46. Adjusted pixel pitch is 1.32 times the minimum pixel pitch that can be achieved by the layout PG3M1. Pixel pitch can be normalized to the minimum pixel pitch of given technology. This normalization factor was called pixel pitch factor. Comparing with the published data on pixel pitch, and fill factor, designed pixel layout, PG3M1, provides better fill factor, pixel capacity and technology scaling.

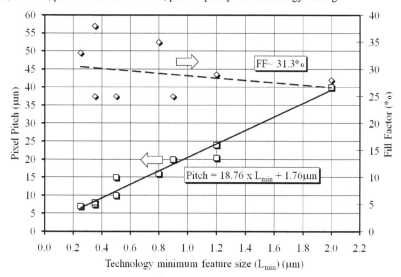

Figure 4.44. Fill factor and pixel pitch of CMOS photogate pixels versus technology minimum feature size of published data between 1990 and 2004, [REFL2].

103

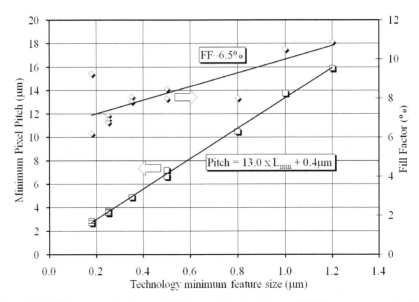

Figure 4.45. Minimum pixel pitch and fill factor of the pixel layout PG3M1 versus the technology minimum feature size.

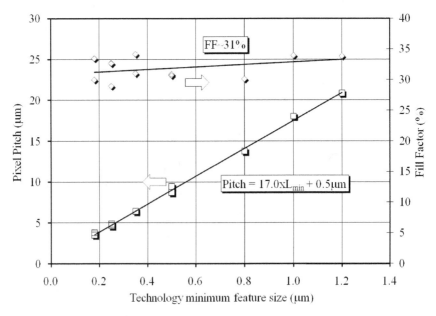

Figure 4.46. Adjusted pixel pitch and fill factor of the pixel layout PG3M1 versus the technology minimum feature size.

4.10.3 Pixel Full-Well Capacity Models

Photogate pixel capacitance versus pixel pitch factor of the photogate pixel layout (PG3M1) for different process technologies are shown in Figure 4.47. One million electrons (1Me-) well capacity equivalent pixel pitch versus the technology minimum feature size is shown in Figure 4.48. It is given with the equation [4.66].

$$\text{Pitch (1Me}^-) = 13.0 \, L_{min} + 3.85 \, \mu m \qquad [4.66]$$

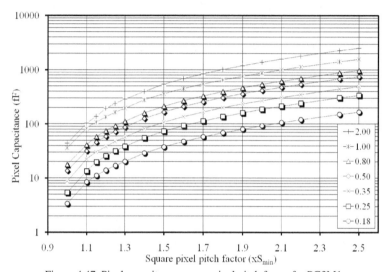

Figure 4.47. Pixel capacitance versus pixel pitch factor for PG3M1.

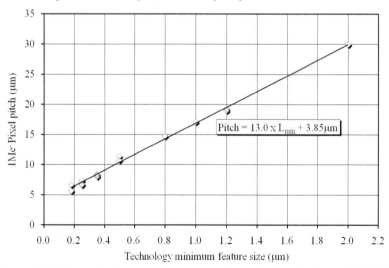

Figure 4.48. 1Me- equivalent pixel pitch versus minimum feature size for pixel PG3M1.

4.11 Pinned Photodiode Type Active Pixel Sensor (PPD APS)

Pinned photodiodes have been used in charge coupled devices (CCDs) since their inception in late 1970s [Hynecek79, Hynecek80]. They have advantages in the area of blue spectrum response, dark current and image lag. For these reasons, pinned photodiodes have been associated with high performance CCD image sensors [Burkey84]. Consequently, with added processing steps, pinned photodiode technology was adapted by CMOS image sensor manufacturers in mid-1990 [Lee95, Lee97]. Besides their known benefits over CCD technology, CMOS image sensors with pinned photodiode have emerged with excellent quantum efficiency (QE), very low dark current and density of white spots. In addition, good modulation transfer function (MTF) and crosstalk would be achieved with appropriate process modifications and pixel design.

Working principle of pinned photodiode is similar to that of the standard photodiode. Differences are n+ layer in standard photodiode is replaced typically by a deep and lightly doped n-type layer that is sandwiched between a p+ surface and a p- epitaxial substrate layers. Cross sections and potential well diagrams of a pinned photodiode pixel are depicted in Figure 4.49. Pinned photodiode holds collected photo generated electrons in a charge pocket below the silicon surface determined by the depth of p+ layer allowing efficient collection of photo electrons generated near the surface. Generated holes below and above the n- depletion regions are swept to the substrate that is connected to ground potential. If the pinning p+ layer dose is high enough, surface potential stays pinned to ground potential leading to reduction in dark current.

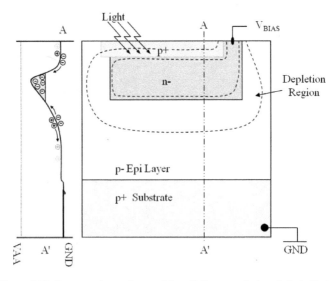

Figure 4.49. Cross section and potential well diagram of pinned photodiode.

106

4.11.1 Three Transistor (3T) Pinned Photodiode APS Pixel

A tree transistor (3T) pinned photodiode APS pixel along with the ideal potential well diagram of the structure are depicted in Figure 4.50. Pixel composes of reset (M1), source follower (M2), and select (M3) transistors like the standard APS pixel. Important design considerations are design of the reset transistor, source follower transistor's connection to the n- layer, and overlap of the p+ pinning layer over the n- layer on the reset transistor side. Depending on the overlap of p+ region on n- layer, reset transistor has imbalanced source and drain diffusion dopings. This makes it difficult to control threshold voltage and it's uniformity over a large image sensor area. Pixel signal is read through the source follower amplifier, and the gate of source follower transistor is connected to the buried n- layer. A special contact should be used to form good connection. This contact should be placed close to the reset gate to increase the pixel fill factor. Depending on the design rules of n- contact, reset transistor size should be adjusted, typically increased. If these issued were not tackled properly, they cause potential barrier in charge pocket. This leads to image lag and increased dark current. It also increases charge feedthrough due to the increased reset transistor size [Inoue99, Inoue03].

Another important design parameter is the pinned potential (Vpin) that depends on the doping level of n- layer. It is typically customized for application in where the imager is used. In ideal case, pin potential (Vpin) is designed lower than the pixel steady state reset level (Vreset).

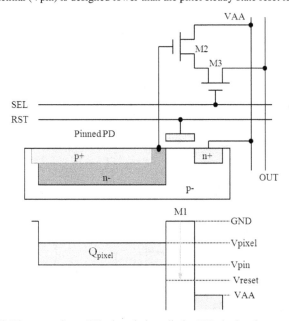

Figure 4.50. Three transistor (3T) pinned photodiode APS pixel and energy diagram.

This leads to a complete cleaning of charges when the reset transistor is turned on, effectively reducing KTC noise. True correlated double sampling (CDS) is possible with the 3T pinned photodiode APS pixel unlike the standard 3T APS pixel counterparts.

4.11.2 Four Transistor (4T) Pinned Photodiode APS Pixel

Four transistor (4T) pinned photodiode composes of pinned photodiode, floating diffusion node, reset transistor (M1), source follower transistor (M2), select transistor (M3), and a transfer transistor (M4). Pixel schematic and ideal potential well diagram is shown in Figure 4.51.

Pinned photodiode composes of the p+/n-/p- sandwich structure. Pinned photodiode region is separated from the floating diffusion node. Collected charges are transferred to the floating diffusion site through a transfer transistor (M4). Floating diffusion node is reset and buffered like typical 3T photodiode APS pixel except the readout sequence of pixel signals were altered. In 4T pinned photodiode APS pixel, first, the floating diffusion node is reset, and sampled on to the column sample and hold capacitance. Next, the pinned photodiode charge is transferred to the floating diffusion node, and this signal is sampled on another column sample and hold capacitance allowing true correlated double sampling (CDS) operation afterwards.

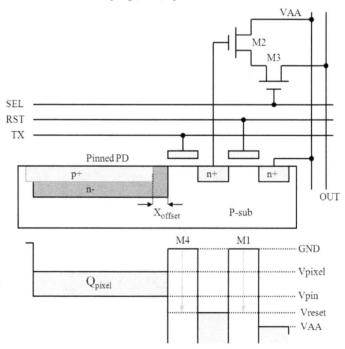

Figure 4.51. Four transistor (4T) pinned photodiode APS pixel and it's energy diagram.

108

Typically, a space (offset) left between the edge of the p+ region and the transfer transistor's gate edge on the photodiode side. Amount of this offset effects the operation of 4T pinned photodiode APS. It has to be optimized during process development. If the offset is large, than an undesired charge pocket appears in photodiode well. If the offset is small or negative than a barrier appears. Barrier cause certain amount of charge to be left in the well from the previous integration period causing image lag. Charge pocket cause the response curve to be nonlinear, and cause image lag, too. Potential well diagram related to the pocket and barrier, as well as the relation between the offset amount and the barrier and the pocket size are shown in Figure 4.52.

Light response of the pixel for ideal and non-ideal cases was illustrated in Figure 4.53. In ideal case the pixel saturates at exposure level depicted as L4. If the pocket exists, it is filled with the electrons first until the signal level reached to the pin potential with smaller capacitance and higher responsivity. Then, the effective pixel capacitance becomes similar as the ideal case. As a result, pixel gets saturated earlier than the ideal case at exposure level of L3. If there is a barrier exists, than the pixel photodiode has to be filled first. Until the barrier equivalent light level (L2) is reached, there would be no effective signal available from the pixel. After barrier reached, pixel response curve follows the ideal curve. Because of the barrier, pixel gets saturated at higher exposure level, L5, but cannot be able to image dark scenes effectively.

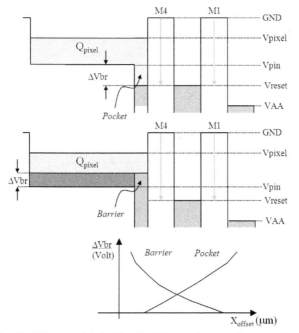

Figure 4.52. Non-idealities associated with offset distance between the p+ region and the transfer gate edge on photodiode side.

109

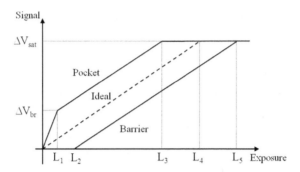

Figure 4.53. Pixel signal versus the exposure.

4.11.3 Shared Floating Diffusion Type Pinned Photodiode APS Pixels

Fill factor is one of the main design considerations when the APS pixel pitch is reduced. In four transistor (4T) pinned photodiodes APS pixels, fill factor is increased by sharing the floating diffusion region by a number of pinned photodiode sites [Guidash00]. This way effectively number of transistors per pixel is reduced. Thus, more silicon area could be used for photodiode area. If one floating diffusion is shared by two adjacent photodiodes (2-way sharing), than the number of transistor per pixel become 2.5 as shown in Figure 4.54. If the number of photodiode sites connected to a common floating diffusion region are four (4-way sharing), than the number of transistors per pixel becomes 1.75.

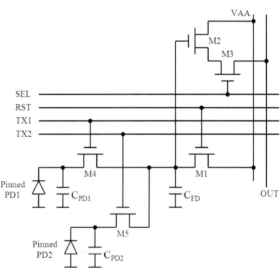

Figure 4.54. Schematic of a 2.5 transistor per pixel photodiode APS with 2-way sharing.

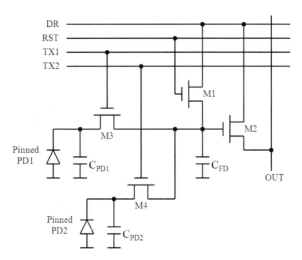

Figure 4.55. Schematic of a 2.0 transistor per pixel photodiode APS with no select transistor and 2-way sharing.

Number of transistor per pixel can further be reduced by both sharing the floating diffusion region, and removing the select transistor (M3) from the pixel site as shown in Figure 4.55 for the case of 2-way sharing of floating diffusion node. 1.5T APS can be achieved with this scheme with 4-way sharing.

4.12   Snap-Shot Type Active Pixel Sensor (SS-ASP)

Snap-shot type active pixel sensor (SS-APS) pixel technologies were developed to overcome the drawbacks of rolling shutter operation of the standard CMOS APS type pixels [Aw96, Yadid-Pecht91, Guidash97, Yang98, Lauxtermann99, Kyrmski99]. They replace the mechanical shutter used in most of the camera systems by allowing simultaneous integration of all pixels in the sensor array. A simple schematic of snap-shot APS pixel is shown in Figure 4.56.

Working principle of the snap-shot active pixel sensor is similar to the pinned photodiode APS pixel. The differences are photosensitive region is a simple photodiode without pinning, and photodiode and floating diffusion regions are reset separately (or together). Floating diffusion node capacitance is used as an analog frame memory to be readout as randomly and as frequently as required.

Transistors M1 and M5 are used to reset the pixel photodiode (PD) and the pixel analog memory (C_{MEM}). Charge is transferred between photodiode and the analog memory through the transfer transistor M4. Analog memory and photodiode reset controls are separate. Analog memory node is read through source follower transistor (M2) when pixel select transistor (M3) is turned on.

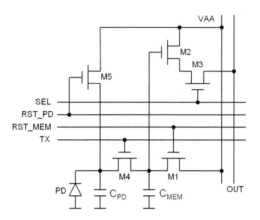

Figure 4.56. Schematic of a snapshot type APS (SS-APS) pixel.

Important design consideration is the holding efficiency of analog memory in snap-shot pixel. It is also called shutter efficiency. Stored pixel value in analog memory node is subject to different interferences and looses reducing shutter efficiency. These are related, how the memory capacitance was build, and what type of process technology was used. Typically, analog memory capacitance is associated with the gate capacitance of the source follower transistor (M2) and the drain/source diffusions of the transfer (M4) and the reset transistors (M1). Since these diffusions respond to impinging photons same way as the photodiode diffusions, they have to be protected from the light. Typically, an overlaying metal layer is used to protect the memory. Another critical issue for shutter efficiency is diffusion of the charges generated outside of the photodiode region. These charges either collected by the analog memory diffusion, or by the photodiode.

4.13   Logarithmic Type Active Pixel Sensors (L-APS)

Logarithmic type CMOS pixels were developed to improve the intra and inter scene dynamic range of the CMOS active pixel sensors [Chamberlain83, Mead85, Mann91] [Ricquier92, Ricquier95, Dierickx96]. Unlike the standard APS pixels that work in integration mode, logarithmic APS pixel works in continuous mode. Thus, maximum frame rate of logarithmic type APS image sensor depends on the time constant of the photosensitive node.

Schematic of a three transistor (3T) logarithmic type APS (L-APS) pixel is shown in Figure 4.57. Pixel composes of a bias transistor (M1), a photosensitive element (PD), select transistor (M3) and a source follower buffer transistor (M2). Only tree signals (pixel select, pixel output, and pixel supply) are routed in the pixel array as oppose to standard 3T PD-type APS pixel of four. Thus the overall fill factor and quantum efficiency of logarithmic pixel is better than that of the standard 3T APS pixel.

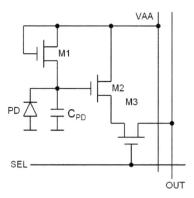

Figure 4.57. Schematic of tree transistor logarithmic type active pixel sensor (APS)

Logarithmic light response was achieved by connecting one or more NMOS transistor (M1) between pixel supply and photosensitive node in serial diode mode. Higher the number of transistor connected in serial, larger the dynamic range, and lower the response time.

Photodiode voltage is self-adjusted to the lighting condition such that the load transistor current equals to photocurrent collected by photodiode. This results in a logarithmic transformation of photo signal and wide intrascene dynamic range. Logarithmic pixel permits true random access in both space and time. Drawbacks to this non-integrating approach include slow response time for low light levels, and large fixed pattern noise. Although, it is able to cover over six orders of magnitude in incident light level, the sensor has a small signal-to-noise ratio (45 dB) due to temporal noise and small voltage swings.

4.14   Other CMOS Pixels

A large array size requires smaller active pixels with a reduced in-pixel component count and routing wires for access control. Examples of such active pixels were introduced by Oba [Oba97] and Ihara [Ihara98]. In these pixels, numbers of control signals were reduced and select and reset transistors were omitted. Pixel is selected and/or reset with appropriate biasing of row signal through in-pixel capacitance.

A photogate CMOS APS with a floating-gate sense amplifier that allows multiple non-destructive, double sampled reads of the same signal was developed at JPL for use with oversampled column-parallel ADCs [Mendis93b].

A floating gate sensor with a simple structure was reported by JPL/Olympus [Nakamura95]. This sensor used a floating gate to collect and sense the photo signal and features a compact pixel layout with complete reset.

There has been significant work on retina-like CMOS sensors with non-linear and adaptive response capabilities. While their utility for electronic image capture has not yet been demonstrated, their very large dynamic range and similarity to the response of the human eye offer intriguing possibilities for on-chip intelligent imaging [Koch 95].

4.15   Summary

It is essential to develop CMOS pixel property models to estimate electrical, optical and physical performance parameters and trends. In this chapter different aspects of modeling were investigated for CMOS pixels.

A modified hybrid threshold voltage equation was derived from BSIM3v3 threshold equations for MOS devices. It was develop to simplify hand calculations of transistors that have minimum feature sizes and varying back gate bias voltages. This model was proved to have a peak-to-peak accuracy of 2% over wide range of back-gate bias, supply voltage, and CMOS processes. Based on this formulation, CMOS APS pixel reset level and pixel source follower signal ranges were derived. Based on the technology model parameters and developed signal range models, pixel capacitance was modeled to estimate pixel full-well capacity.

A tool called pixel physical characteristics table (PPCT) was developed for characterizing physical and electrical properties of CMOS pixel layout design. Parameter trends were determined for CMOS pixels utilizing PPCT. They were developed to be used during technology evaluation and pixel design. APS pixel pitch versus the technology minimum feature size was determined for 1Me- well capacity. Other CMOS pixel technologies were reviewed as well in this chapter without extensive modeling.

CHAPTER 5 PROTOTYPE CMOS IMAGE SENSOR DESIGN

In this chapter, prototype CMOS image sensor design is discussed. Circuit and timing details of the imager were given. Developed prototype development imagers were used for testing ideas of new pixel technologies aimed for improving pixel full-well capacity, and dynamic range. These new pixels are discussed in next chapters.

5.1 Prototype CMOS Image Sensor

There are many ways to test CMOS imaging pixels using test vehicles. Some uses product grade imager platforms to test not only the performance of the imaging pixels, but also their performance in final product environment. Some uses very small array of dumb pixels to measure basic characteristics of the pixel under investigation. One of the known architecture is called fully flexible open architecture (FFOA) that composes of sample and hold circuits, correlated double sampling (CDS) and differential delta sampling (DDS) circuits, and source follower amplifiers [Nxon96, Mendis97a]. Simple FFOA architecture gives very reliable and predictable signal path characteristics. It also allows multiple pixel types with different sizes to be integrated on the same chip. Here we are going to give details of proposed pseudo-differential, series FFOA architecture. Top level block diagram s of the architecture is shown in Figure 5.1.

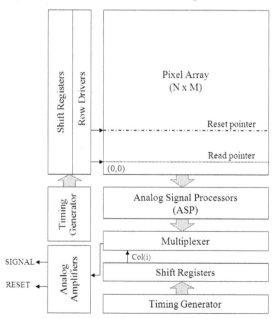

Figure 5.1. CMOS FFOA image sensor block diagram.

115

Proposed CMOS FFOA imager platform composes of an N x M CMOS APS pixel array, row timing generators for row shift registers to set reset and read pointers on the pixel array, row drivers, column analog signal processors (ASPs) containing column buffers, offset cancellation and sample and hold (S/H) circuits, column shift registers controlled by column clock and timing generator, global charge and sample and hold (S/H) amplifiers, and analog/digital buffers.

5.2 Analog Signal Chain (ASC) Topology

Analog signal chain (ASC) composes of circuits between sensing node of pixel or photodiode (PD) and analog output ports of the imager chip. Most of the analog processing such as programmable gain, offset correction, level shifting, correlated double sampling (CDS), differential delta sampling (DDS) and driving is done before the column signals were sent to the global analog processing circuits.

ASC gets it's input from selected pixel. Pixel composes of the photo conversion site or the photodiode (PD), and part of the pixel buffer amplifier. An n-type CMOS common-drain amplifier or source follower is used as pixel buffer amplifier. It is called pixel source follower (PSF). PSF buffers photodiode signals to column sample and hold (CSH) circuitry.

There are tree gain stages exist in proposed pseudo-differential, series FFOA architecture as shown in the Figure 5.2. Pixel source follower (PSF) is the first gain stage and has a typical gain of less that unity. Analog signal processor (ASP) on each column contains another amplifier stage that conditions signals from PSF and CSH. Because of the noise and speed concerns, a source follower amplifier is used. It is called column source follower (CSF). CSF buffers and isolates the CSH from the correlated double sampling (CDS) and delta double sampling (DDS) blocks. PSF and CSF gain stages exist on each column. Thus, column to column matching of these circuits is critical.

Programmable gain amplifier (PGA) and the output analog buffer are shared by all the columns. Column signals are transferred to the global analog amplifiers sequentially pulsing column select switches (CSEL) by the column decoder block.

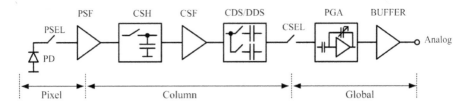

Figure 5.2. Analog signal chain (ASC) from pixel node to chip outputs of prototype CMOS imager.

116

5.3 Analog Signal Chain (ASC) Types

ASC section boundaries and the location of analog signal processing operations depend on the imager architecture. Three major type of ASC architecture is used in CMOS APS imagers; serial, column-parallel, and pixel-parallel.

If all the analog signal processing including programmable gain amplifier (PGA) and ADC operations are done in the pixel boundaries, than the architecture is called pixel-parallel ASC (PP-ASC) architecture. PP-ASC architecture is typically employed in high-speed, low to mid-resolution CMOS imagers. Power consumption, ADC resolution, and pixel sizes are the major design considerations of PP-ASC imagers.

ASC architecture that shares PGA and ADC or analog buffer blocks between columns is called serial ASC (S-ASC) architecture. In S-ASC, column of pixels shares PSF, and CSH circuits. CSH contents are shifted out sequentially to shared PGA block, and digitized by ADC block before it is sent off-chip. Depending on the imaging array size, and frame rate requirements, global readout section may require to operate at very high throughput rates. This requirement can be relaxed by employing more than one channel of PGA and ADC blocks on the signal path. High speed ADC architectures, such as pipelined, folding, and flash are typically employed in S-ASC architectures.

If entire signal processing operations are done in each column, than the ASC architecture called column parallel ASC (CP-ASC) architecture. In column parallel architecture speed and power requirements are relaxed for PGA and ADC blocks. Thus low speed, compact ADCs such as single/dual slope, successive approximation, and sigma-delta can be used.

5.4 Analog Signal Chain (ASC) Design Issues

Speed, linearity, power consumption, and noise of CMOS APS imager can be optimized by addressing critical design issues of ASC circuits. In this section these design issues are going to be address to achieve scientific grade analog signal chain design.

5.4.1   Pixel Source Follower (PSF)

Pixel signals are first buffered through pixel source follower (PSF) amplifier. Typical PSF amplifier composes of four NMOS transistors as shown in Figure 5.3. Two of the PSF transistors are placed in the pixel; amplifying source follower transistor (M1), and pixel select transistor (M2). The other two are placed at the bottom of the pixel array shared by number of pixel rows. They are the column enable transistor (M3) and the source follower load transistor (M4). Transistors M2, and M3 are used as switch. M2 is used for accessing the pixel in 2D array, while M3 is used for

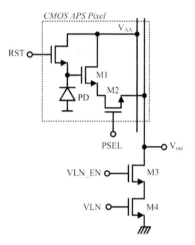

Figure 5.3. Pixel source follower (PSF) amplifier.

enabling and disabling the PSF during pixel and column readout periods, respectively, to save power.

In true correlated double sampling (CDS) pixel readout operation, pixel is reset to a known voltage first, and this photodiode (PD) voltage is buffered through the PSF to the CSH block. After the integration period, integrated signal level is buffered ans sampled in CSH block. In between these two signal levels, PD node swings from a known reset level, typically the supply voltage, to a lower voltage that is proportional to the light level integrated on the photosensitive node. In dark this difference is small, in bright it is large.

In CMOS APS imagers, typically an n-type PSF amplifier is used. N-type PSF could buffer high input signals with better gain, and linearity than that of the lower signals levels assuming that the pixel source follower sinking current is kept same. For constant PSF current, gain and linearity roll off resulting in gain variation and the non-linearity between the buffered signals. Thus, the column transistor size and the bias conditions are arranged such a way that this roll off is minimum in 1volt effective pixel voltages. Effective PSF output voltage can be written in terms of PSF gains while buffering the PD reset voltage and PSF gain while buffering the PS signal voltage as ;

$$\Delta V_{out}(PSF) = PSF\_Gain(reset) * V_{PD}(reset) - Gain(signal) * V_{PD}(signal) - V_{offset} \qquad [5.1]$$

Simulated PSF gain and gain variation versus the effective PSF output voltage is shown in the Figure 5.4 a, and b. In this particular simulation a 5Volt, 0.5μm CMOS process model parameters were used. PSF has better gain and gain linearity at lower PSF bias currents as shown in the simulation results. Major drawback of reducing the PSF current are reduced gain bandwidth and readout speed resulting in higher 1/f noise.

118

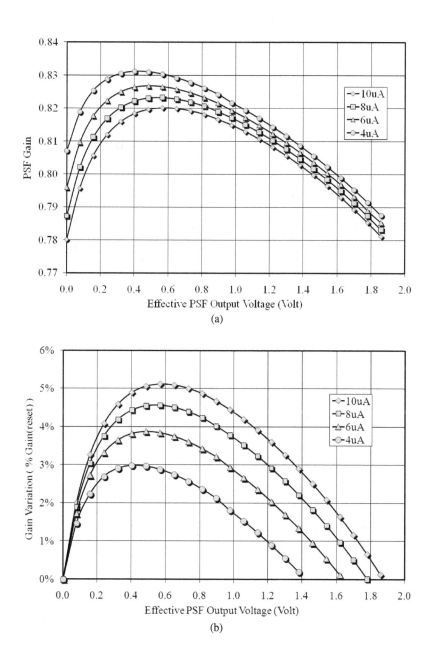

Figure 5.4. Effective PSF output voltage versus a) PSF Gain, b) PSF gain variation normalized by the PSF gain while buffering photodiode's reset signal level.

Size of the amplifying transistor, M1, of the PSF is critical. Since it is located in the pixel, it should not to occupy large pixel area thus might have minimum size. However, this results in poor PSF gain, column-to-column PSF gain variation and eventually high fixed pattern noise (FPN). Thus, the device size of M1 has to be optimized for specific process technology to achieve better matching, noise and parasitics.

Channel length of M1 is typically chosen to have longer than minimum size to increase PSF gain and uniformity. Another consideration with choosing the device size of M1 is the extra capacitive loading on the PD node. The gate-source and the gate-body capacitances are the two dominant non-linear capacitances that adds extra loads to the PD node. This extra capacitance could change the pixel PD node capacitance and is proportional the device size of M1, the PSF gain, and the PD node voltage. These capacitances can be written as;

$$C_{par}=Gain(PSF)*C_{gs}(M1)+C_{gb}(M1) \qquad [5.2]$$

$$C_{gs}(M1)=C_{gso}*W(M1) \qquad [5.3]$$

$$C_{gb}(M1)=C_{ox}*W(M1)*L(M1) \qquad [5.4]$$

Thus the total pixel capacitance during integration periode is smaller than that of the pixel readout periode. During integration pixel capacitance is equal to;

$$C_{pix,it}=C_{pd}+C_{gb}`(M1) \qquad [5.5]$$

$$C_{pix,read}=C_{pd}+Gain(PSF)*C_{gs}(M1)+C_{gb}(M1) \qquad [5.6]$$

$C_{gb}`(M1)$ is smaller than the $C_{gb}(M1)$ because during integration time select transistor, M2, is turned off removing the channel charge of the M1, and leading to lower oxide capacitance (C_{ox}) and gate to body capacitance. Parasitic capacitance from the transistor M1 become significant if the pixel PD node capacitance has to be small for high sensitivity.

Another effect is related to noise of PSF related to the size of M1. Larger the W*L multiplication factor, lower the column to column gain mismatch and lower the FPN noise. Also 1/f noise reduces with larger W*L factor.

5.4.1.2   *Sizing Pixel Select Transistor (M2)*

Since the M2 is used as switch, minimum channel length could be used. Channel width on the other hand might be smaller than that of the M1. Especially, common column bus side of the source diffusion causes extra capacitive loading on this bus. Size of this diffusion has to be minimized to reduce the PSF loading in large format imagers. Also, source diffusion need to be protected from light by an overlaying metal layer to prevent unwanted signal discharges in CSH blocks in certain ASC architectures.

5.4.1.3 Sizing the PSF Load Transistor (M4)

Load transistor (M4) sets the amount of current used in PSF. Column-to-column matching of PSF gains and currents depend on column-to-column matching of load transistors located at the bottom of the pixel array. Thus it is very important to choose device size to achieve best parameter matching for M4.

Longer width and length result in better device matching in CMOS technologies. Thus, a long and wide NMOS device has to be used for M4. Matching can further be improved by reducing the column bias current. Reduced current also increases the PSF gain and the linear range as shown in Figure 5.5. It is because the overdrive voltage (Vgt=Vgs-Vth) of M4 can be reduced by reducing PSF current that leads lower Vds voltage in where M4 goes out of saturation. One important effect for long and wide M4 is that the extra loading created on the common bias line, VLN.

5.4.1.4 PSF Bias Driver

The load transistor, M4, is driven by a simple current mirror structure while enable transistor, M3, is typically controlled digitally in CMOS APS imagers. For better matching and mirroring, M times the column current run through the driver transistor M5 that has M times the device size of the CSF load transistor M4. Typically, driver current, Idrv, is generated by an on-chip current DAC that could be controlled digitally as shown in Figure 5.6. An extra transistor, M6, can be included on

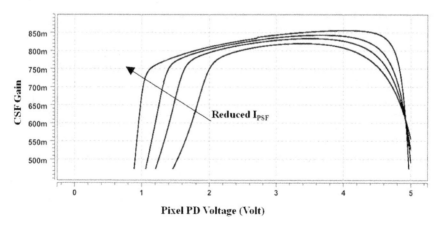

Figure 5.5. Pixel PD node voltage (VFD) versus PSF gain for different I_{PSF} currents.

121

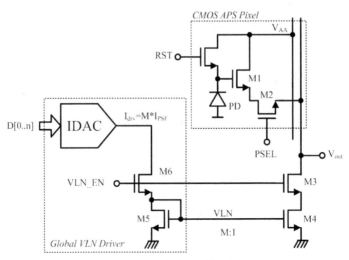

Figure 5.6. Standard PSF bias driver.

driver side to power down the driver after the pixel read. It is also possible to drive enable transistor, M3, with an analog signal. This allows cascode amplification and reduces the noise contribution of the M4 improving minimum detectable pixel signal in CMOS APS imager with the expense of reduces linear range on the PSF.

5.4.2 Column Sample-and-Hold (CSH)

Column sample-and-hold (CSH) is placed in each column for sampling pixel signal and reset levels buffered by pixel source follower (PSF). Typically, it composes of one or more sampling capacitances, and a number of switches. Depend on the ASC architecture front-end sampling could be done in charge mode or in voltage mode in CSH block.

5.4.2.1   Voltage Mode CSH Circuits

In voltage mode front-end sampling, signals are buffered by column amplifier(s) and level shifted for larger dynamic range operation in the subsequent stages. Parametric variations of these amplifiers are corrected either by local (per column) or global correction techniques such as delta double sampling (DDS). Two voltage mode CSH circuits are shown in Figure 5.7. CSH in Figure 5.7a have two identical circuits composing of a switch, a hold capacitance, and a buffer. Since there are two buffers, matching and column FPN associated with the CSH is poor in the first circuit. Column FPN is improved by adding extra switches and using single buffer in Figure 5.7b. In second circuit, both sampled pixel reset and signal voltages passes through same CSH buffer improving column FPN. However, timing and biasing gets more complex in second CSH circuit.

122

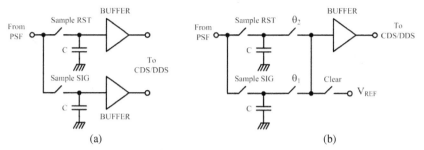

Figure 5.7. Voltage mode CSH circuits.

5.4.2.2   Charge Mode CSH Circuits

Two charge mode CSH circuits are shown in the Figure 5.8. In charge mode CSH blocks, CDS and DDS could be integrated within the CSH block and sent effective pixel voltage to the global amplifiers. First circuit in Figure 5.8a does not have any active amplifiers and composes of identical switches and capacitors. It provides simple charge mode sampling and averaging. It has good column FPN performance as long as the switches and capacitors are well matched column-to-column. Second circuit composes of an amplifier, two capacitors and switches. Since it has an active amplifier circuit on each column, FPN has the main concern in this topology. Because of the amplifier, early gain can be implemented while achieving correlated double sampling operation. If gain is implemented, overall ASC readout noise floor could be reduced with the second topology.

5.4.3   Correlated Double Sampling (CDS)

A differential output correlated double sampling (CDS) circuit for voltage mode CSH is shown in Figure 5.9. It composes of two identical capacitors and multiple of switches to generate absolute pixel signal and level shift this effective voltage for global amplifier block. For voltage mode, delta double sampling (DDS) operation could be achieved by shorting two sampling capacitances in CSH block. CDS operation can be implemented in charge mode CSH blocks. Thus, charge mode CSH blocks achieve better form factor and used in commercial CMOS imagers.

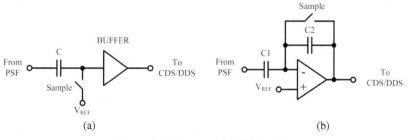

Figure 5.8. Charge mode CSH circuits.

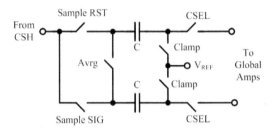

Figure 5.9. CDS circuit for voltage mode CSH.

5.5 Address Decoders and Drivers

Decoders are an integral part of CMOS image sensors. They are used to address column and rows on the pixel array and column processing circuits. Row decoders are used for selecting row of pixels on the pixel array during pixel sampling time in where pixel signals are sampled on to column sampling circuits. Column decoders are used for selecting the sampled column signals sequentially. Row decoder operated at much slower speed than the column decoders. Thus speed requirements for row decoders are more relaxed than that of the columns.

5.5.1 Decoders

Many different decoders are available in CMOS technology to choose from. Fundamentally, two type of addressing scheme is possible. They are the binary and gray decoding schemes. They both generate the same output while their inputs are somewhat different. Major difference is that the only one bit changes from one address input to the next in gray decoder while as much as (n-1) bits could change in binary decoders. That means in gray decoders, only one bit changes at both input and output digital words. As a result for a large decoder, such as 2048 or 1024-bit, total number or charged or discharge node is only two in gray decoders, and n in binary decoders. This gives advantage of low glitch operation for gray decoders over the binary counterparts.

In terms of physical logic design, there are more than two possible design choices. Most commonly used ones are; NAND/NOR gate based multiple input decoders, Pre-charge logic based decoders, Dynamic logic gate based decoders. All these circuits allow random address generation. If the addressing is sequential, shift register based decoders might be an alternative to these circuits. Unlike the logic gate base decoders, shift register based decoders produce less glitch between transitions. Especially high speed column decoding operation, they become advisable considering very sequential readout of column APSs of CMOS image sensors. Thus, for both row and column decoding, shift register based decoders were used in prototype CMOS image sensor architecture.

There are two popular way to implement shift register based decoder. One uses static shift registers, and the other uses dynamic one. For high speed decoding applications, such as column address decoding, dynamic shift registers are preferable. It is because the retention time of shift register content is very short in the order of tens of nanoseconds. For decoding operation that require long retention times in the order of tens of microseconds, static shift registers has to be utilized. Two such circuit realizations are shown in Figure 5.10. Both circuit uses single clock and could be reset globally. Dynamic shift register is much more compact in terms of layout composing of 8 CMOS transistors, while static shift register contains 26 CMOS transistors. Scalability is the main advantage of shift register based decoders. Without changing anything, same core shift register can be tiled as many as needed to generate address signal. Logic gate based decoders need to be re designed if the address is increased.

5.5.2 Row Decoder and Driver

Row decoder composes of the core addressing element, a shift register in this case, and couple of AND gates to generate row select (SEL), and pixel reset (RST) signals as shown in Figure 5.11. If reset and select signals are not boosted, no other than strong digital buffers are

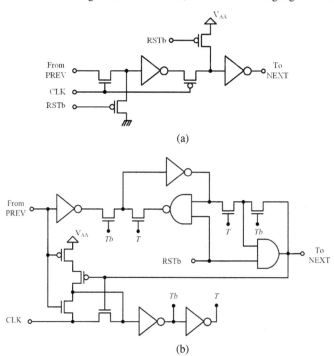

(a)

(b)

Figure 5.10. Single clock shift register circuits. a). Static, b). Dynamic.

125

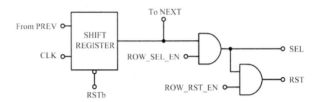

Figure 5.11. Shift register based row decoder.

required to drive RST and SEL signals to the pixel array. Shift register output is directly connected the the input of the next shift register. Row decoder in Figure 5.11 assumes read and reset pointer is the same on the pixel array. If a very short integration times are required in the application, two of this circuit has to be used; one for read pointer addressing, and the other is for reset pointer addressing.

5.5.3 Column Decoder

Column decoder works much higher frequencies than that of the row decoders in CMOS image sensors. Column decoder and drivers not only generate column select signal, but also generates CDS, DDS, and other control signals. Accurate generation of these signals are typically effects the image quality and column related FPN. Thus, extra attention has to be given on generating timing signals of the column decoder and drivers.

5.6 Timing and Controller Generators

Timing signal generators for the row and column decoders and drivers shares the same circuits. They are shown in Figure 5.13. Clock generator is a non-overlapping pulse generator. MCLK signal could be either row or column clock signal. Reset and shutter timing generator is used for initializing the shift registers. There are two of them used in row block to control reset and read pointer on the pixel array. It provides initial signal for the first shift register, and does generate continuous rolling shutter operation by using last shift register output and rolling shutter enable (ROLL_EN) signals.

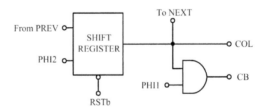

Figure 5.12. Shift register based column decoder.

126

Global amplifier timing signals are generated along with the column decoder signal as shown in the Figure 5.13c. It is because; column readout has to be coordinated with the global programmable gain amplifiers. Global amplifier reset signal and output sample and hold signals are the two critical signal. Shift register reset and shutter control block is used for both column and row shift registers.

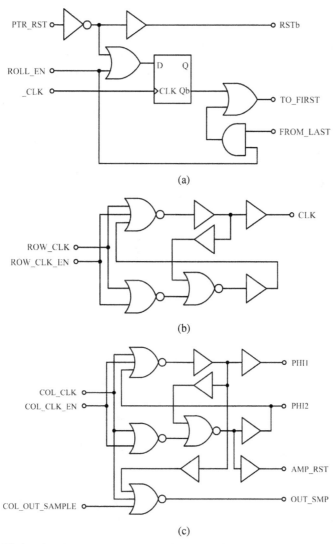

(a)

(b)

(c)

Figure 5.13. Timing signal generators; a) reset and shutter timing generator b) row clock generator, c) column decoder and global amplifier timing generator.

127

5.7 Prototype CMOS Image Sensor Design

5.7.1 Architecture and Control Signals

A single channel, pseudo-differential, series analog output APS architecture was chosen with rolling shutter capability as depicted in Figure 5.14. General design issues were discussed in previous sections. Not only each block but also each control signal is marked on Figure 5.14. These signal names are used in following sections.

Shift registers were used for row and column address decoding. They were preferred because of the scalability and stitchability of shift register circuits, and sequential readout nature of the column analog signal processor (ASP) blocks. Two separate row clock generators were used for row shift registers to set row read and row reset pointers on the pixel array allowing rolling shutter imaging operation. This also allows very short integration periods during imaging. Column analog signal processor (ASP) contains column buffers, offset cancellation and sample-and-hold (S/H) circuits. Column shift registers, pseudo-differential global charge amplifiers, and output sample-and-hold (S/H) amplifiers were controlled by column clock and timing generator. Analog and digital buffers were used for driving external analog and digital signals to the internal circuits.

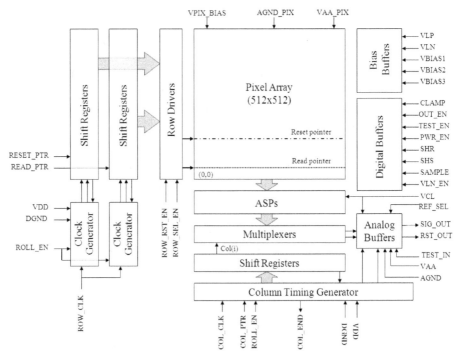

Figure 5.14. Block diagram of the CMOS APS prototype chip.

128

Pixel arrays were divided into sub-arrays containing multiple reference and test pixels. Three versions of the prototype image sensor were designed based on the same architecture, but different pixel sizes and array formats. In first prototype version (PROTO1_1), pixel pitch was set to 18 μm. Pixel array size was 424 by 424. It was divided into 16 sub-arrays with 106 by 106 pixels. In second prototype version (PROTO1_2), pixel pitch was set to 15 μm. Array size was 512 by 512. It was divided into 6 sub-arrays with 256 by 170 pixels. In third prototype (PROTO1_3), pixel pitch was set to 15μm, and single three transistor (3T) CMOS APS pixel was used. Array size was 512 by 512. [Ay01]

5.7.2 Analog Signal Chain (ASC) Design

5.7.2.1 Analog Signal Processor (ASP) Design

Analog signal processor (ASP) composes of half of the pixel source follower (PSF), a column sample-and-hold (CSH) circuit, correlated double sampling (CDS), differential delta sampling (DDS) circuits, and number of switches and capacitances as shown in Figure 5.15.

PSF composes of four NMOS transistors. Upper half of the PSF is located in the pixels while the lower half of it is shared by one column of pixels that were connected to the common column bus. Two of the transistors work as switch. One is located in the pixel. It is activated by row decoder during pixel read time (SEL$_i$). The other one is located at the bottom of column bus. It is turned on during pixel read periode, and turned off after that to save power.

CSH composes of a mixed voltage mode and charge mode sample and hold circuitry. Voltage mode CSH is placed first followed by a charge mode CSH. Voltage mode CSH is used just for buffering and level shifting the PSF output. It composes of a switch transistor controlled by SAMPLE signal and a sampling capacitor. Voltage mode CSH also contains an analog test input circuitry for characterizing analog signal chain gain from PSF to the chip outputs. PSF signal is buffered through a PMOS type source follower amplifier, isolating voltage mode CSH from proceeding block. It also increases sampling speed of the ASP, and extends linear operation range of the imager. Charge mode CSH composes of two sampling capacitors and few control and select transistors. CDS and DDS operation is done on charge mode CSH section to reduce fixed pattern noise (FPN) associated with the PSF and CSH source follower. During DDS operation, voltage domain sampled signal in first CSH is converted into charge domain signal. Pixel reset and signal voltages were sampled on two capacitances with respect to the clamp voltage (VCL) and CDS operation is performed during readout.

ASP control signals were generated by column shift registers and column timing generator circuits. Column shift registers selects the column ASPs sequentially during column readout period.

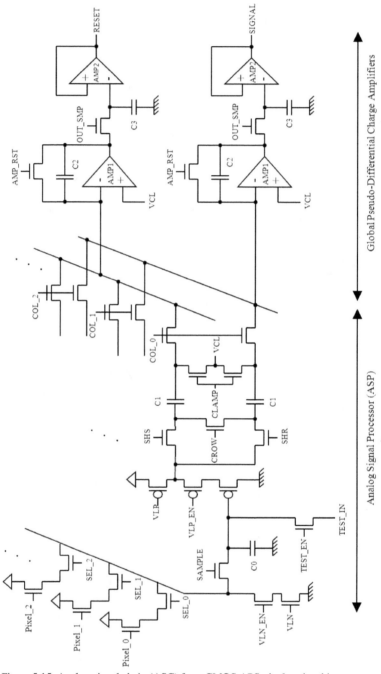

Figure 5.15. Analog signal chain (ASC) from CMOS APS pixel to the chip outputs.

5.7.2.2   Global Charge Amplifier (GCA) Design

Sampled pixel reset and signal levels in ASPs were driven off chip through two identical global charge amplifier (PGA) channels providing pseudo-differential analog outputs. Each composes of one charge amplifier (AMP1), one sample and hold (S/H) circuit and a unity gain analog buffer (AMP2) as shown in Figure 5.15. Control signals of these amplifiers are generated by column timing generator as shown in Figure 5.13c.

Two types of operational transconductance amplifiers (OTAs) were used in GCA. First stage amplifier (AMP1) is a load compensated cascode operational transconductance amplifier (OTA) with p-type input stage as shown in Figure 5.16. Cascode transistors were used at the output stage to improve systematic offset, power supply rejection ratio performances, open loop DC gain, and to build a more symmetrical amplifier structure which is very important in pseudo-differential signal channels. Current mirroring ratio between input branch and the output cascading branch was chosen to be unity to optimize the input referred noise performance of the amplifier. P-type input differential pair was designed to achieve at least 2 volt peak-to-peak input level around clamp voltage (VCL) with high gain and 60° phase margin. 2 pF load capacitance and less than 25 ns settling time to 10-bit accuracy were other design specification for the first amplifier. Supply voltage was 5 volt and power consumption was targeted less than 10mWatts during design.

A p-type input stage was also used in the second stage amplifier (AMP2). This amplifier has a compensated output stage to drive large capacitive load. Bandwidth, gain, setting time, and input range of the second stage amplifier was similar to the first stage amplifier.

Simulated characteristics of both amplifiers are shown in Table 5.1.

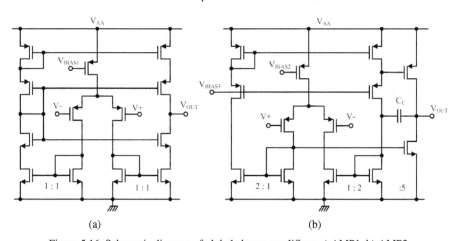

Figure 5.16. Schematic diagram of global charge amplifiers; a) AMP1, b) AMP2

Table 5.1. Simulation results of global charge amplifiers (GCAs) at process corners.

	AMP1	AMP2	
Supply Voltage	5	5	Volt
Input Range	1.00-3.25	1.10-3.60	Volt
Absolute Input Range	2.25	2.5	Volt
Gain (Max.)	70	73	dB
Gain (@Input range)	>61	>65	dB
Gain Bandwidth	85	70	MHz
Phase Margin	60	58	Degree
Load	2	15	pF
Settling (@1Vpp input)	<20	<25	ns
Power Consumption	<10	<35	mWatts

5.7.3   Prototype Image Sensor Timing

5.7.3.1   Chip Timing

Typical sequence of processing an imaging sensor array starts with scene integration. Usually, pixels in 2D array do not start integrating simultaneously but rather in a sequential manner. Start of integration periods of two adjacent rows is the minimum integration time that an image sensor achieves. Reading a row starts with storing a row of pixels information in column sample-and-hold circuits. This period is called row time. Usually, gain and signal conditioning performed in this period. After row time, each column is scanned out by addressing each analog signal processor (ASP) and transferring the stored signals through the common global amplifiers. This period is called column time. It is determined by the speed of global amplifiers, and number of columns in the array. In addition, a step of conditioning is performed on a separate row to start integration for that row. Thus, row selection logic drives 2 separate sets of addresses. The first address set is used for reading addressed row. It is called read address pointer. The second address is the reset address pointer used for conditioning a separate row to start integrating. The time lapse between starting the integration for a particular row and reading that row is called integration time. During the integration time, photodiode node voltage drops proportional to the light level on the pixel. All these were depicted for the prototype pixel in Figure 5.17.

Integration time adjustment based on two-address pointer scheme is depicted for 3 row and full frame integration time in Figure 5.18 for the prototype sensor architecture. Integration time can be adjusted by setting number of rows to be reset before row read operation starts. Depending on the integration time set by user, reset pointer advances one row at a time. After the set integration time is reached, read address pointer is initialized and row read operation performed.

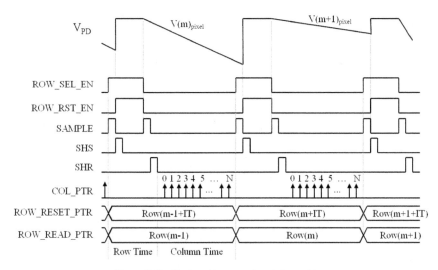

Figure 5.17. Timing diagram of prototype chip.

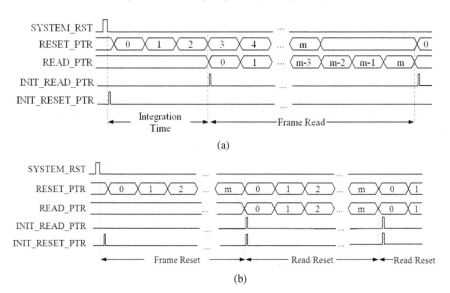

Figure 5.18. Timing arrangement for a) row base integration, b) full frame integration.

Prototype chip digital input signals during row initialization and row read periods are shown in Figure 5.19. Row initialization and integration time adjustment periods compose of three sub-sections. First sub-section is the overhead in where row shift registers are initialized. Second is the integration time (IT) sub-section that is repeated number of times to advance reset address pointer on the image sensor array. During each of these sub-sections, row of pixels were reset by enabling

133

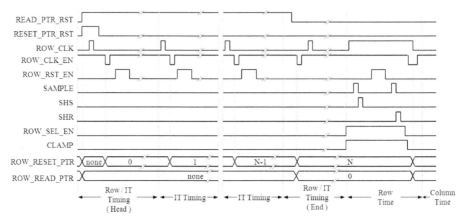

Figure 5.19. Row initialization and read time periods

the ROW_RST_EN signal. IT sub-section should be as wide as the row and column times combined. After reset pointer advances to an address point where the integration time was set, a third sub-section begin to initialize read pointer. Right after the last sub-section completed, row time period starts. In this period, a row pointed by read pointer is selected and sampled onto column analog signal processor while another row pointed by the reset pointer is reset.

5.7.3.2   Row and Column Read Timing

Right after the row read time period has ended, column time period starts as depicted in Figure 5.20 for the prototype chip architecture. Column time period composes of two sub-sections. First one is the column initialization period. During this time period, column shift registers were initialized. The second sub-section is the column read period in where all sampled signals were shifted out of column analog signal processors (ASP) to the global charge amplifiers. ASP contents shifted out one by one to the output port through the global charge amplifiers and buffers. Timing diagram for single column and control signal for the global amplifiers are shown in Figure 5.21.

5.7.3.3   Global Amplifier Timing

ASP outputs (Vsignal and Vreset) were amplified on a reference clamp voltage (VCL), sampled on output sample and hold capacitance, and buffered off chip by a unity gain amplifier. Control signals were generated on chip by a non-overlapping clock generator. ASP outputs sampled during low period of the column clock (COL_CLK) signal, and hold during the high period. Amount of signal on the clamp level for reset and signal outputs are equal but with the opposite sign.

134

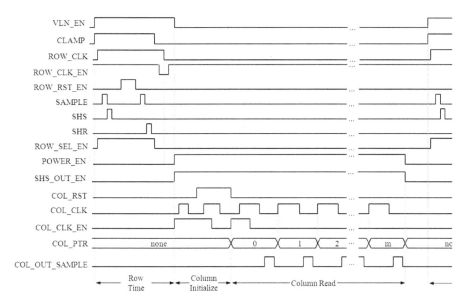

Figure 5.20. Column time period

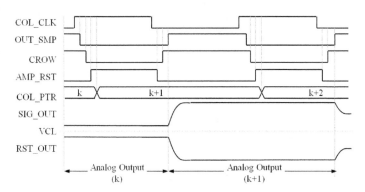

Figure 5.21. Global amplifier control signals

5.8 Prototype CMOS Image Sensor Specifications

Three prototype CMOS APS image sensors were designed and fabricated. Prototypes were sharing same circuits and pads except pixels. First prototype CMOS imager (PROTO1_1) composes of 424 by 424 pixel array containing 16 different CMOS APS pixels. Thus, pixel array was sub-divided into 106 by 106 pixels. Pixel pitch of these pixels was 18μm. Different pixel ideas were tried on the first prototype. Second prototype CMOS APS imager (PROTO1_2) composes of a single photodiode type CMOS APS pixel. Array size was 512 by 512. The pixel pitch was 15μm.

The third prototype CMOS imager (PROTO1_3) also composes of 512 by 512 pixel array containing with 6 different CMOS APS pixel designs. Pixel array was sub-divided into 170 by 256 pixels. Pixel pitch of these pixels was 15µm. Specifications of the CMOS APS chips were given in Table 5.2.

Microphotographs of the CMOS APS prototype chips are shown in Figure 5.22. They were fabricated in double-poly, triple-metal (2P3M), 5volt, 0.5 µm CMOS process. Prototype chips were designed to fit in 84 pins, 0.47 mil cavity PGA package (open lid PGA84L). Total die area of the prototype CMOS APS imager was 95 mm^2 (9.75mm x 9.75mm).

Table 5.2. CMOS APS prototype image sensor specifications

Process	0.5 µm CMOS (2P3M)
Array Dimension	424 x 424 (PROTO1_1) 512 x 512 (PROTO1_2) 512 x 512 (PROTO1_3)
Pixel Size	18 µm x 18 µm (PROTO1_1) 15 µm x 15 µm (PROTO1_2) 15 µm x 15 µm (PROTO1_3)
Pixel Types	3T, Photodiode type, CMOS APS 3T Hybrid Photodiode-Photogate type, CMOS APS
Output Format	Differential Analog
Frame Rate	>30 fps
Power Consumption	<200 mW @ 5.0 V
Power Supply	5.0 Volt
Package	PGA 84L
Die Size	9.75 mm x 9.75 mm

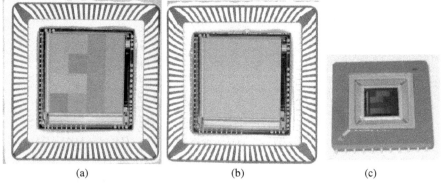

(a)   (b)   (c)

Figure 5.22. Microphotographs of CMOS APS prototype image sensors; a) PROTO1_1,
b). PROTO1_2, c) packaged PROTO1_1.

5.9 Characterization System Design

A custom test board was designed to test the prototype imagers. A PC user interface program was written for collection and analyzing image data. Test board accommodates different ways of PC interfaces including PC frame grabber, PC parallel port, and SVGA video output. It had 12-bit analog-to-digital converter (ADC), frame memory (FIFO), programmable clock generator, an FPGA, and bias and reference generator. FPGA was used for control signal generation for both test chip (DUT) and other blocks. Block diagram of the test board is shown in Figure 5.23a. Pictures of the test board are shown in Figure 5.23b.

Both analog frame grabber and PC parallel port interfaces were used for characterization of the prototype chips. A characterization program was developed for parallel port interface of PC. Program communicates with the test board through parallel port at speeds up to 1Msample/sec. Image acquisition, gamma correction, regional image statistics, zooming, file operations, and other characterization features were included in the user interface design. Following measurements were performed; noise performance, light saturation, dynamic range, quantum efficiency (QE), dark current, sensitivity, and conversion gain.

5.10   Measurement Results

Measurement results of the PROTO1_2 CMOS APS imager with photodiode type 3T pixel is reported in this section. Measurement results of the other imagers are reported in the next chapters.

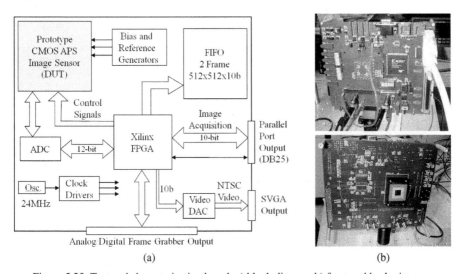

(a)   (b)

Figure 5.23. Test and characterization board; a) block diagram b) front and back pictures.

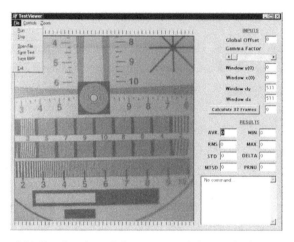

Figure 5.24. User interface of viewer, test, and characterization software.

5.10.1 Analog Signal Chain (ASC) Measurements

Analog signal chain (ASC) gain from pixel photodiode to analog chip outputs were measured utilizing both measurement and simulation techniques.

Using the simulation tools, pixel source follower (PSF) input-output characteristics was determined first. The transfer function was determined with a 2th order linear fit function from PSF output (V_{COL}) to the input (V_{PD}) as shown in Figure 5.25. Transfer function equation was determined beyond PSF or ASC input output range for characterizing signal path with good accuracy. PSF transfer equation was determined in terms of PSF output. Because, all signals are referred back to PSF input. Also test input voltage is applied at the output node of the PSF during signal sampling, and this voltage has to be referred to the PSF input to determine effective pixel voltage versus the ASC gain. From simulation, it was determined pixel photodiode reset level was 4.27 volt while the PSF output is 2.65Volt for this reset voltage. These were used for determining the effective PSF output voltage.

Enabling the test switch in column sample and hold block (see Figure 5.15), a signal voltage is sampled on CSH block. This voltage is subtracted from the pixel photodiode voltage. During ASC characterization single row was selected and pixel photodiode is kept at reset state. Test input was scanned between 0 and 5 volt. It is subtracted from buffered pixel reset level in each column CDS/DDS blocks and passed through the rest of the ASC blocks. Effective pixel voltage versus the measured ASC output is shown in Figure 5.26. Effective pixel voltage was determined by referring test input voltage to the PSF input by using the equation shown in Figure 5.25 and subtracting it from the pixel photodiode reset voltage which was 4.27 volt. Ratio of the differential output voltage

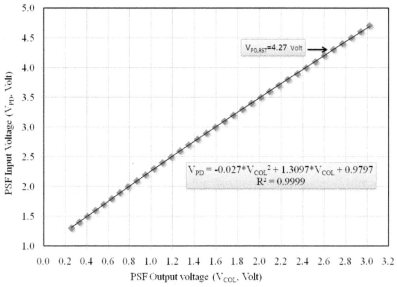

Figure 5.25. Simulated input output characteristics of PSF.

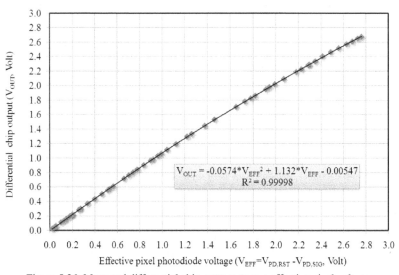

Figure 5.26. Measured differential chip outputs versus effective pixel voltage.

versus the effective pixel photodiode voltage is the gain of the analog signal chain from the pixel photodiode to the chip output. This gain was designed to be unity to make chip characterization simple. Measured ASC gain is shown from PSF output to chip output (upper) and pixel photodiode node to chip output (lower) in Figure 5.27. During design global charge amplifier gain was set to

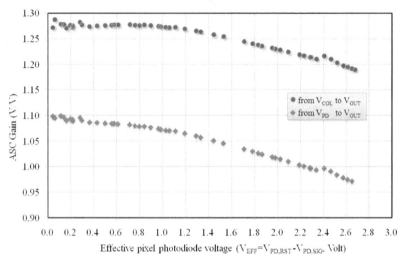

Figure 5.27. Measured analog signal chain (ASC) gain versus effective pixel voltage.

1.5 V/V by choosing C1/C2 ratio in Figure 5.15. Upper plot in Figure 5.27 include this gain and the column P-type source follower amplifiers gain. Thus ASC gain from test input to chip output is the multiplication of the global charge amplifier gain and the column source follower gain. Gain of the column p-type source follower was calculated to be around 0.85 V/V. ASC effective linear range was close to 2.7 volt with gain of close to unity as shown in Figure 5.27.

Using the ASC transfer function equation in Figure 5.26, all measurements at the outputs were referred back to the pixel photodiode node. ASC gain could amplify effective pixel voltages

5.10.2 Dark Current Measurement

Dark current can be specified as a number of input referred electrons generated per second in a pixel, or as a current per photodiode area. It is measured by plotting average pixel output voltages at different integration times while the imager placed in a dark and controlled temperature environment. The slope of the plot is used for calculating the dark current. If the conversion gain and pixel pitch of the imager is known, dark current can be converted to ampere per squared area, or to electron per second.

Dark current of the prototype imager (PROTO1_2) was measured. Chip output was measured and referred back to the pixel photodiode node by using the ASC transfer function, and plotted for different integration time as shown in the Figure 5.28. Measured dark current was 32.75 mVolt/sec or 4889e-/sec.

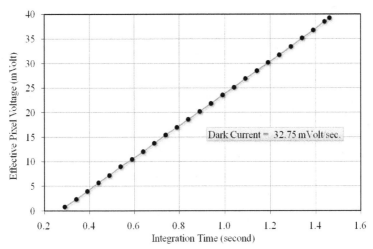

Figure 5.28. Measured dark current of the pixel in PROTO1_2 imager.

5.10.3 Conversion Gain and Full-Well Capacity Measurements

Conversion gain measurement is done by stepping a light source from complete darkness to full-well illumination in precisely measured increments. At each illumination level, at least 60 frames were captured at 65msec integration time and mean and variance of each pixel were computed by following the procedure given in Section 2.2.6. Conversion gain is then computed by measuring the slope of the photon transfer curve in the photon shot noise limited region. Photon transfer curve is shown in Figure 5.29.

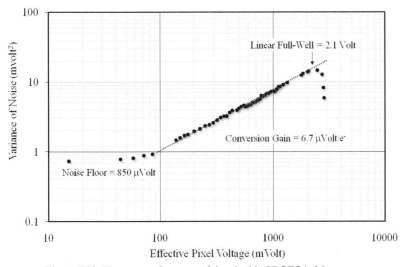

Figure 5.29. Photon transfer curve of the pixel in PROTO1_2 imager.

141

Measured conversion gain of the PROTO1_2 pixel was 6.7 µVolt/electron with full-well saturation voltage of 2.1 Volt. Thus, full-well capacity of the pixel was about 315 Ke⁻.

5.10.4 Dynamic Range and Signal to Noise Ratio Measurements

Measured noise floor of the imager was 0.85 mvolt or 127 electrons. Dynamic range of the imager was calculated to be 2480 or close to 68dB at pixel linear full-well range of 2.1 Volt or 315K e⁻. However, imager ASC could achieve larger than 70dB dynamic range including nonlinear response range of the image sensor as shown in Figure 5.30.

Signal to noise ratio (SNR) was also measured as shown in Figure 5.31. Larger than 55dB SNR was achieved at linear full-well range of the pixel.

Fixed pattern noise (FPN) was measured as plotted with respect to the effective pixel voltage in Figure 5.32. FPN level of less than 0.5% was achieved over 80% of the linear full-well.

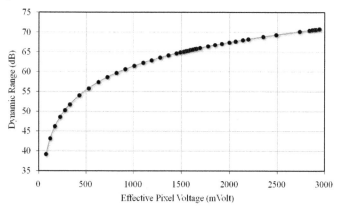

Figure 5.30. Dynamic range measurement results.

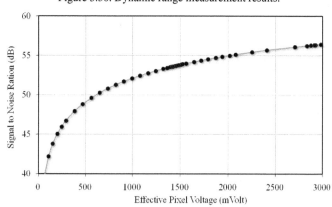

Figure 5.31. Signal to noise ratio (SNR) measurement results.

142

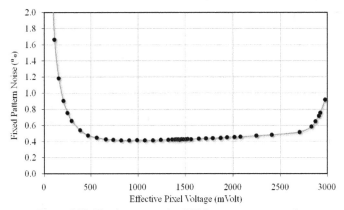

Figure 5.32. Fixed pattern noise (FPN) measurement results.

5.10.5 Linearity and Sensitivity Measurements

Sensitivity of the pixels was measured by stepping a light source from complete darkness to full-well illumination in precisely measured increments. At each illumination level the mean values of the differential chip outputs were calculated and recorded. A sharp pass band green filter was used at the output of the light source. Light level is measured in foot/candle, and converted to lux. Measurements were performed at 175 msec integration time. For each pixel, mean pixel value in volt versus exposure in lux*sec were plotted and the sensitivity of the pixels were calculated by determining the slope of the plots below pixel full-well capacity of 2.1 Volt. Measured sensitivity of the PROTO1_2 imager is shown in Figure 5.33. Measured sensitivity was close to 2.0 Volt/Lux.sec.

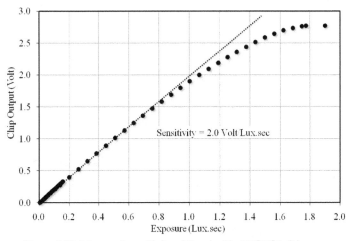

Figure 5.33. Measured sensitivity of the pixel in PROTO1_2 imager.

143

5.10.6 Quantum Efficiency (QE) Measurements

Quantum efficiency (QE) of the PROTO1_2 pixel was measured between 390 nm to 700 nm with 10nm steps and between 700nm and 1100nm with 100nm steps as shown in Figure 5.34. Peak QE was about 31% at 580nm and 500nm.

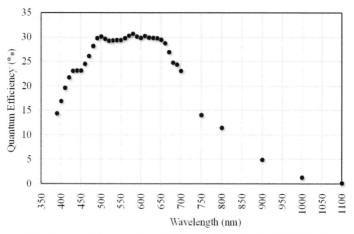

Figure 5.34. Measured pixel quantum efficiency of the pixel in PROTO1_2 imager..

5.10.7 Other Measurements

Other measurements such as power consumption under different bias conditions, imager characterization at different supply voltages other than 5.0V nominal value, maximum frame rate, and imager noise at different timing conditions were performed. Maximum full-frame rate for PROTO1_2 was determined to be around 40 FPS with power consumption of 196mWatts on 5.0Volt power supply.

Since all timing and control signal are same for all three prototype CMOS image sensors, only pictures taken by some of them are shown in Figure 5.35 and Figure 5.36 as example.

5.11 Summary

In this chapter a prototype CMOS image sensor platform design was discussed. Designed platform was used as test and characterization vehicle for CMOS APS pixels. Not only electrical but also photonic characterizations of prototype imagers were reported. One of the imager (PROTO1_2) was fully characterized and tabulated in Table 5.3. Pixels in other prototype imagers are reported in next chapters.

Figure 5.35. Picture taken by prototype CMOS APS image sensor, PROTO1_3.

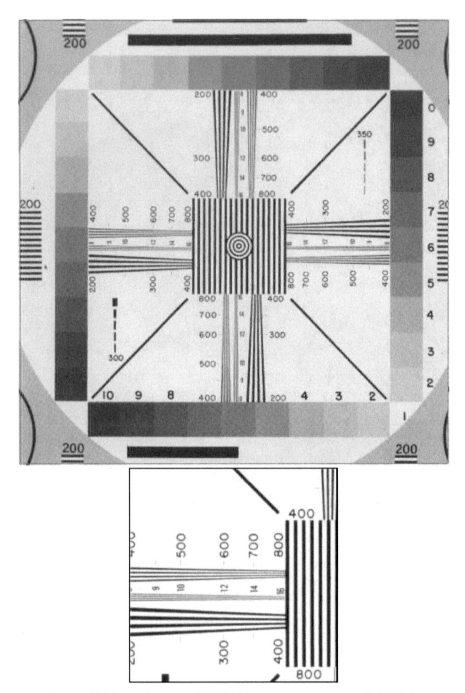

Figure 5.36. Picture taken by the CMOS APS prototype image sensor, PROTO1_2.

146

Table 5.3. Measurement results of prototype CMOS image sensor, PROTO1_2.

Technology	0.5 μm CMOS (2P3M)
Imager ID	PROTO1_2
Array Size	512 x 512
Pixel Pitch	15 μm
Pixel Type	Photodiode APS
Pixel Fill Factor	75%
Output Format	Differential Analog
Frame Rate (up to)	40FPS
ASC Gain	1.05 V/V
Responsivity at 550 nm	2 Volt/Lux.sec
Noise Floor	850 μVolt
Linear Full-Well Capacity	315,000 e-
Linear Full-Well Voltage (Vsat)	2.10 Volt
Conversion Gain	6.7 μVolt/e-
Fixed Pattern Noise (FPN)	< 0.5% (@80% of Vsat)
Signal-to-Noise Ratio (SNR)	55 dB (@Vsat)
Dynamic Range	68 dB (@Vsat)
Dark Signal	< 33 mVolt/sec
	< 5000 e-/sec
PRNU	< 0.23% (of Vsat)
Quantum Efficiency	31% @ 580nm (max.)
	29% @550nm
Power Supply	5 Volt
Power Consumption	< 200 mWatts
Package	PGA 84L
Chip Size	9.75 mm x 9.75 mm

CHAPTER 6 LARGE FULL-WELL CAPACITY CMOS PIXEL DESIGN

Large pixel full-well capacity or well depth is required for wide dynamic range image sensing that is one of the main requirements for scientific image sensors. Full-well capacity of a photodiode type APS pixel is simply calculated by multiplying total junction capacitance and maximum measurable voltage swing of the photodiode. It is given in section 2.2.5 with equation [2.18]. However, noise in pixel and readout circuits could limit the dynamic range because pixel dynamic range is defined by the ratio of maximum number of charge that could be held in a pixel to noise floor as given in equation [2.20]. Dynamic range is as plotted for different full-well capacity and noise floors in Figure 6.1. Consequently, wide dynamic range operation could be achieved by increasing pixel full-well capacity and pixel voltage swing and by reducing noise floor of pixel and readout circuits.

In this chapter, investigation results of different pixel full-well capacity improvement methods for CMOS APS pixels were discusses. Two new methods were proposed. These are photodiode peripheral utilization method (PPUM), and in-pixel capacitance utilization method that uses a new hybrid photodiode-photogate (HPDPG) pixel structure. Proposed two methods were tried on silicon, and measurement results are reported, as well as the test imager architecture, timing, and the circuits.

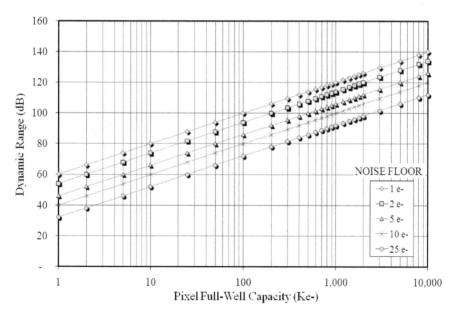

Figure 6.1. Dynamic range versus pixel full-well capacity and noise floor.

6.1 Pixel Full-Well Capacity Improvement Methods

Pixel full-well capacity can be increased by increasing total junction capacitance of photo conversion site in where photon to charge conversion takes place. In photodiode type, tree transistor (3T) APS pixels (which will be dealt with throughout this chapter), this could be achieved by increasing photodiode area or its doping concentration. Improvement is also possible by connecting other types of capacitors available in CMOS process to the photo conversion node.

Increasing photodiode area may not be a preferred way to increase pixel full-well capacity. Because, it increases pixel size and total die size, resulting in fabrication yield to drop. In addition, forming large pixel arrays become costly in terms of optics that is going to be used in the imaging system.

It is possible to increase pixel full-well capacity by connecting a high density, linear capacitor to photodiode node if it is available in the process. Three types of capacitors could be used in the pixel; poly-insulator-poly (PIP) capacitor, metal-insulator-metal (MIM) capacitor, and poly-insulator-channel (PIC) capacitor.

PIC type of capacitor is formed by using standard NMOS or PMOS type of active devices. It is also called metal-oxide-semiconductor capacitor or MOSCAP. For linear operation, channel of the MOSCAP device has to be formed by properly biasing gate, drain, and source terminals. This biasing requirement, generally, limits pixel signal and linear operation range of the pixel. MOSCAP provides largest unit area capacitance values than any other linear capacitor available in sub-micron CMOS processes today. They also are sensitive to light, improving not only effective pixel full-well capacity but also optical characteristics of pixel.

PIP and MIM kind of capacitors are light insensitive but provides relatively lower unit area capacitance values. PIP and MIM capacitors do not have any restriction on the way they connected to the photosensitive or photo conversion regions. Thus, they do not limit the pixel signal range like the MOSCAPs. However, typically, minimum physical size that can be drawn is prohibitively limited by the process design rules.

It is possible to increase photodiode capacitance without increasing the pixel size. This is achieved by increasing the total peripheral of the photodiode by opening number of holes on the photodiode region. This is called photodiode peripheral utilization method (PPUM) and will be discussed in next sections with measurement results, [AY081]. The other method is to use a DC biased MOSCAP in parallel with a photodiode to improve and expand not only the pixel full-well capacity, but also pixel linear light response range. This new pixel is called hybrid photogate-photodiode APS pixel and is reported in next sections with measurement results [AY082].

6.2 Photodiode Peripheral Utilization Method (PPUM)

Theory behind the photodiode peripheral utilization method (PPUM) is that, if the pixel pitch is restricted to a certain size, then pixel full-well capacity could be increased by opening holes in the photodiode's diffusion area. These diffusion holes could be used to increase photodiode parasitic capacitance, by increasing the perimeter capacitance of the photodiode region in certain process technologies as shown in the Figure 4.17. Holes also can increase spectral response of a photodiode by utilizing lateral collection of charges converted close to the semiconductor surface [Fossum99, LeeJ01] at the edges of photodiode.

A reverse-biased PN-junction diode is used in photodiode (PD) type CMOS APS pixels as a photon conversion and charge (electron) storage element. Total capacitance of a photodiode diffusion defines key pixel performance parameters. For example, wide-dynamic-range pixels require large pixel full-well capacity and low readout noise. Photodiode full-well capacity is comprised of two components: bottom plate (area) and side wall (peripheral) junction parasitic capacitance. Pixel designer controls the size of the photodiode diffusion area or the bottom plate, while diffusion junction depth and doping concentration are process and technology dependent. The unit area junction capacitance (C_A) and unit peripheral junction capacitance (C_P) of photodiode diffusion are given with equations [6.1] and [6.2]. They include technology and design parameters, for the first-order capacitance that contributes to total full-well capacity of a photodiode pixel.

$$C_{PD} = C_A \times A + C_P \times P \tag{6.1}$$

$$C_{PD} = \frac{C_{J0A} \cdot A}{\left[1 - \dfrac{V_{PD}}{\Phi_B} \right]^{M_J}} + \frac{C_{J0SW} \cdot P}{\left[1 - \dfrac{V_{PD}}{\Phi_{BSW}} \right]^{M_{JSW}}} \tag{6.2}$$

where

C_A, C_P	unit area and peripheral junction capacitance,
C_{J0A}, C_{J0SW}	unit zero-bias area and peripheral junction capacitances,
A, P	area and peripheral of the photodiode regions,
Φ_B, Φ_{BSW}	built-in potential of area and side-wall junctions,
M_J, M_{JSW}	junction grading coefficients of area and side-wall junctions,
V_{PD}	photodiode junction voltage.

Other parasitic capacitances due to the reset and readout transistors in pixel contributing to total photodiode junction capacitance are shown in Figure 6.2 for a three-transistor (3T) PD-APS pixel. These parasitic capacitances contribute to total pixel capacitance differently in different modes of pixel operation. Right after photodiode reset and during scene integration periods, overlap capacitances C_{O1} and C_{O2} and gate-to-body capacitance of the readout transistor M2 (C_{B2}) add to the total photodiode capacitance. During a readout period, miller capacitance C_{M2} and overlap capacitances C_{O1} and C_{O2} contribute to the total photodiode capacitance. Contribution of pixel circuit parasitic capacitances is described by the following equations during imaging (equation [6.3]) and readout (equation [6.4]):

$$C_{par,imaging} = \left[W_{M1} \cdot L_{OL,M1} + W_{M2} \cdot \left[L_{M2} + L_{OL,M2} \right] \right] \cdot C_{ox} \qquad [6.3]$$

$$C_{par,read} = \left[\frac{2}{3} \cdot W_{M2} \cdot \left[L_{M2} - 2 \cdot L_{OL,M2} \right] \cdot \left[1 - G \right] \right] \cdot C_{OX} + \left[W_{M1} \cdot L_{OL,M1} + W_{M2} \cdot L_{OL,M2} \cdot \left[2 - G \right] \right] \cdot C_{OX} \qquad [6.4]$$

where

W_{M1}, W_{M2}	channel width of the reset and source-follower transistors,
$L_{OL,M1}$, $L_{OL,M2}$	channel overlap length of the reset and source-follower transistors,
C_{OX}	unit gate oxide capacitance,
G	pixel source follower gain factor.

C_A and C_P of a few CMOS process technologies, with minimum feature sizes between 2.0μm and 0.18μm is shown in Figure 4.17. Unit-area capacitance is larger for deep sub-micron devices with a minimum feature size less than 0.5μm, due to the increased channel-stop doping-level (for better device isolation, higher diffusion doping concentrations, and shallower junction depths). Thus, peripheral junction capacitance could be better utilized in processes that have equal or more unit peripheral junction capacitances than in processes with less than 0.5μm minimum feature size, by opening holes in the photodiode region. As will be shown in the next sections, this will not only improves the total full-well capacity of the pixel, but also improves the spectral response of it in short wavelength photons.

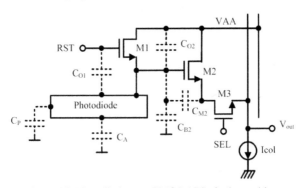

Figure 6.2. Photodiode type CMOS APS pixel parasitics.

6.2.1 Photodiode Lateral Collection Improvement

Photosensitive element in APS pixels, photodiode (PD), works in charge integration-mode where pixels are accessed at the end of integration time period. When it is accessed, photodiode is read and then cleared for next scene integration. Figure 6.3 shows the cross-section of a PN-junction photodiode formed in a CMOS process. The photodiode is reverse-biased and formed by using the shallow N+ doped, drain-source diffusion of an NMOS device. A bias voltage applied to the N+ region forms a depletion region around the metallurgical PN-junction, which is free of any charge because of the electrical field. Any electron-hole pairs generated in this region see the electrical field as shown in the AA′ cross-section view of the photodiode in Figure 6.3.

Electrons "slide" in the opposite direction of the electric field (toward the N+ region), while holes slide toward the P-region; electrons are collected in a charge pocket in the N+ region, while holes are recombined. This type of photodiode has been used in CMOS and early CCD-type image sensors as a photo conversion and collection element.

There are two dark-current issues associated with using the standard N+ diffusion layer of an NMOS transistor as a photosensitive element. First is the dark current induced by stress centers around the N+ diffusion; these stress centers are formed during the field oxide (FOX) formation in standard sub-μm CMOS processes. The second issue is the surface-related dark current generated from the work function difference between the N+ diffusion surface and overlaying isolation oxide layer. This second dark current causes surface recombination centers and defects.

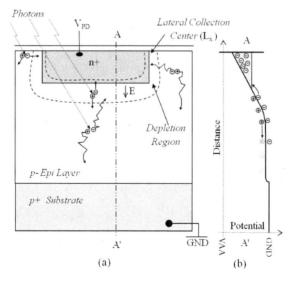

Figure 6.3. Photodiode type CMOS APS pixel; a) cross-section and b) potential-well diagram.

Both localities and stress centers absorb photo-generated electron-hole pairs close to the surface, resulting in quantum loss at shorter wavelengths. As a result, silicon photodiodes shows less sensitivity in the blue region of the spectrum (<400nm), adding the very short absorption depth of these photons. Most blue photons are collected through lateral diffusion of the carriers generated on, or in the vicinity of, a photodiode peripheral—known as peripheral photoresponse or lateral photocurrent [LeeJ03]. Thus, increasing lateral collection centers or peripheral length of a photodiode potentially improves collection efficiency for short-wavelength photons [Fossum99, LeeJ01] as it is shown in Figure 6.4. This method was adopted for UV photodiode devices in P-well CMOS processes [Ghazi00].

6.2.2 Reference Pixel (REF1) Design

A tree transistor (3T) CMOS APS reference pixel (REF1) was designed to normalize measurement results of the test pixels. Layout of reference pixel is shown in Figure 6.5. Pixel pitch was 18μm as the other pixel and integrated in same prototype CMOS APS imager (PROTO1_1). A circular looking photodiode was designed to reduce overall dark current. Row select and reset signals were drawn on top of each other using horizontal metal-2 and metal-3 lines, while the metal-1 was used on vertical direction for pixel output and the pixel supply signals.

The reference APS pixel's photodiode (PD) area and peripheral were 141.7 μm^2 and 44.6 μm, respectively. Unit area and peripheral capacitance of the n+ diffusion layer were calculated by using model equations [4.8], [4.16], [4.25], [4.30], and [4.34]. They were 0.25 fF/μm^2 and 0.22 fF/μm, respectively. Total pixel capacitance was calculated by using equation [4.35] including the miller contribution of the source follower transistor (M2). Miller contribution to the total photodiode capacitance considering source follower gain of 0.75 was calculated to be 1.1fF.

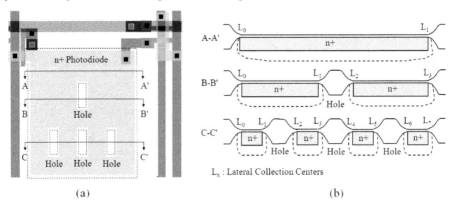

(a)   (b)

Figure 6.4. Improving lateral collection efficiency by photodiode peripheral increase for blue photons; a) Photodiode APS pixel with holes, b) cross-sections of the photodiode.

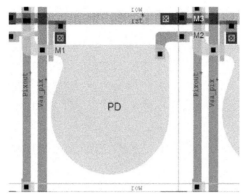

Figure 6.5. 18μm square, 3T CMOS APS reference pixel (REF1) layout.

Peripheral junction capacitance made up of 20% of the total photodiode capacitance. Calculated photodiode capacitance was about 46.5fF.

6.2.3   Measurement Results of the Reference Pixel (REF1)

Dark current was measured at room temperature. It was 10.63 mVolt per second as shown in Figure 6.6. This equals to 3155 e-/sec with the measured conversion gain of 3.37 μVolt per electrons as shown. Measured photon transfer curve of the reference pixel is shown in Figure 6.7. Total measured pixel capacitance was 47.54 fF as oppose to the calculated one of 46.5fF during design. With 1.714 volt effective linear photodiode voltage, pixel full-well capacity was 508Ke-. With same effective linear photodiode voltage, this was calculated to be 498Ke-.

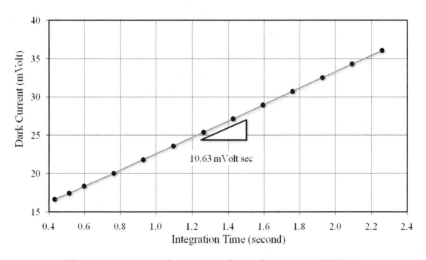

Figure 6.6. Measured dark current of the reference pixel (REF1).

155

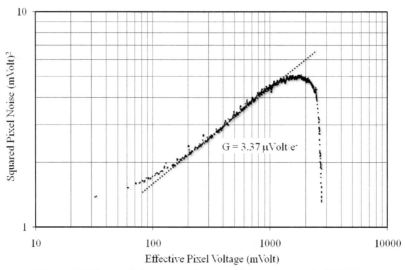

Figure 6.7. Measured photon transfer curve of the reference pixel (REF1).

Measured light sensitivity was 2.44 Volt/Lux*sec as shown in Figure 6.8. Measured peak quantum efficiency was 48.55 percent at 500nm as shown in Figure 6.9. At 400nm, quantum efficiency of the reference pixel was 23.4 percent. Dynamic range of the reference pixel was around 66.4 dB because of the higher noise floor measured in this particular chip. Rest of the measurement and calculations are 4 are listed in Table 6.1 for the reference pixel (REF1).

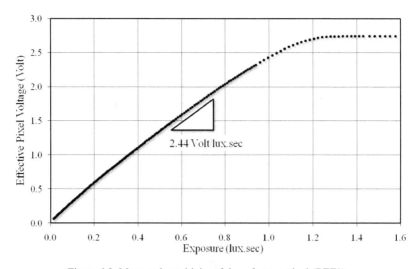

Figure 6.8. Measured sensitivity of the reference pixel (REF1).

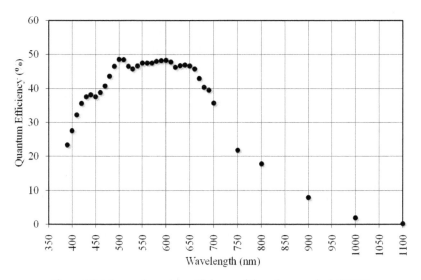

Figure 6.9. Measured quantum efficiency of the reference pixel (REF1).

Table 6.1. Calculated and measured parameters of the reference pixel (REF1).

Parameter	Measured	Calculated	Unit
Sensitivity	2.44		Volt/lux.sec
Light Saturation	1.07		lux.sec
Saturation Voltage	2.74		Volt
Quantum Efficieny	23.41		at 390nm
	47.44		at 550nm
	48.55		peak
Conversion Gain	3.370	3.446	μV/e-
Full Well	1.714	1.714	Volt
	508,736	497,530	e-
Pixel Capacitance	47.54	46.49	fF
Dark Current	10.63		mVolt/sec
	3155		e-/sec
Dynamic Range	66.39	79.67	dB

6.2.4   Test Pixels with Circular Openings

Four test pixels with a number of circular openings were designed to model the peripheral utilization effect on the pixel performances and on the pixel full-well capacity. Layouts of pixels are shown in Figure 6.10. Pixels have 1.6 μm diameter circular openings on their photodiode sites. Total number of openings were 17, 14, 11, and 7 for the pixel layouts called c17, c14, c11, and c7, respectively. Openings were placed on the reference pixel photodiode diffusion in random fashion. Circular openings were chosen to reduce stress centers related dark current in the photodiode area.

157

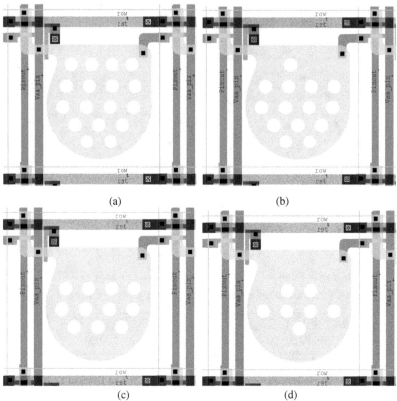

Figure 6.10. Test pixels with circular openings; a) c17, b) c14, c) c11, d) c7.

6.2.5 Measurement Results of the Test Pixels with Circular Openings

All test pixels were placed in the same imager, (PROTO1_1), to compare performance under common imaging and environmental conditions.

6.2.5.1 Conversion Gain and Full-Well Capacity

Pixel full-well capacity was measured through measuring the conversion gain and the full-well saturation voltages of the reference and the test pixels. Measurement results of both pixel conversion gain and the pixel full-well capacity are shown in Figure 6.11. As calculated with the photodiode capacity model given in the chapter 4, pixel full-well capacity increases with proper utilization of pixel photodiode peripheral by using the openings on the photodiode region. Conversion gain of the test pixels were reduced and pixel full-well capacities were increased when more and more photodiode diffusion hole are placed on the reference pixel.

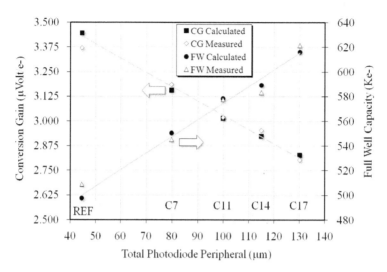

Figure 6.11. Conversion gain and the well depth of the test and reference pixels.

A linear correlation between the total peripheral and the pixel capacity was found. It was because; the unit area capacitance (CA) is almost equal to the unit peripheral capacity (CP) in the technology that the test pixels and prototype CMOS image sensor were fabricated. The area loss was compensated by the peripheral increase by a factor of 2.5. It is because the radius of the opening was set to 0.8 μm, and the opening peripheral was (p=2πr) 5.02655 μm while the area was (a=πr^2) 2.01062 μm^2. Factor of four could easily be achieved by choosing the radius of the openings around 0.5 μm. However, reduced diameter results in depletion region overlap, and lowers peripheral capacitance and utilization. It was also found that the models developed in chapter 4 predicts pixel photodiode capacitance and pixel full-well capacity accurately with better than 2% error.

6.2.5.2   Quantum Efficiencies (QE)

Quantum efficiencies (QE) of the reference (REF1) and test pixels were measured by using the method described in the section 2.5.10. Measurement was performed between 390nm and 700nm with 10nm steps. Measurement results for reference (REF1) and test pixels (c17, c14, c11) are shown in Figure 6.12. In the figure, QE difference between the reference pixel and a test pixel with 17 openings (c17) normalized by the reference QE, was also plotted. Less than 12% spectral response improvement was observed with increased number of openings on the photodiode. Best improvement was achieved at shorter wavelengths and pixel with large number of openings as shown in Figure 6.12 . This trend is more visible in Figure 6.13.

159

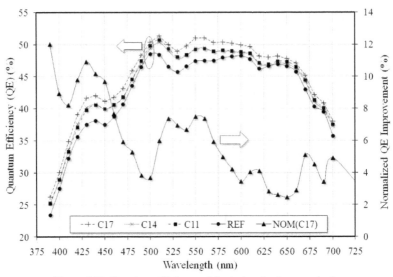

Figure 6.12. Quantum efficiency of the test and reference pixels.

Blue photons that were generate electron-hole pairs close to surface of silicon were collected better laterally at close surroundings of the photodiode area. By adding circular openings, these lateral collection areas were increased. This leads to a better QE response at shorter wavelengths. For deep penetrating photons, collection probability was not increase as much as the surface photons. This leads poorer improvement in longer wavelength as shown in Figure 6.13.

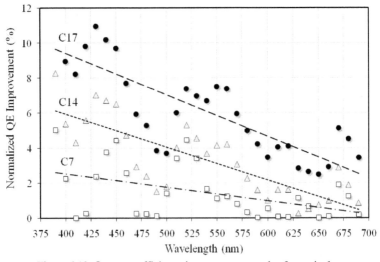

Figure 6.13. Quantum efficiency improvement trends of test pixels.

160

6.2.5.3   Dark Current

As expected, more dark currents were observed from the test pixels that have longer peripherals than that of the reference pixel as shown in Figure 6.14. Dark current in terms of electrons per second increases about one-third of the reference dark electrons when the photodiode peripheral doubles (assuming, the surface dark current effect was neglected). Relation between dark current and the peripheral is exponential. In reality, measured dark current have two components; surface dark current and stress centers related dark current. Surface dark current is related to the area of the photodiode region while the one associated with the stress centers is related to the peripheral. Opening hole on photodiode region reduces surface contribution and increases the peripheral contribution on the total dark current. It is possible to determine the contribution of these two components of the dark current by designing fixed area and varying peripheral test pixels.

6.2.5.4   Linearity and Sensitivity

Sensitivity of the test and the reference pixels were measured by using very sharp green (550 ±20 nm) bandpass filter and 175msec integration time. Measurement results are shown in Figure 6.15 and Figure 6.16. Sensitivity relation with the pixel capacity was extracted by fixing the light wavelength, pixel fill factor, and the integration times. Observed photodiode sensitivity was inversely related to the pixel capacity. Higher the pixel capacity, lover the sensitivity is. Twenty percent increase in pixel capacitance results in seventeen percent decrease in pixel sensitivity as seen between reference pixel (REF1) and the test pixel with 17 openings (C17) in Figure 6.15. Sensitivity relation is more like a second order polynomial.

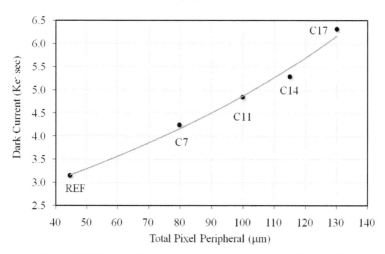

Figure 6.14. Measured dark current rates of reference and test pixels.

161

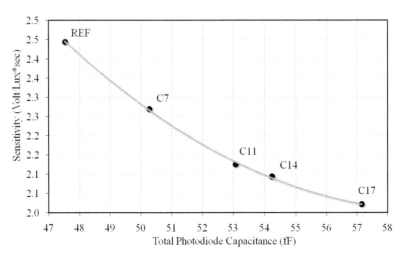

Figure 6.15. Measured sensitivity of the reference and test pixels.

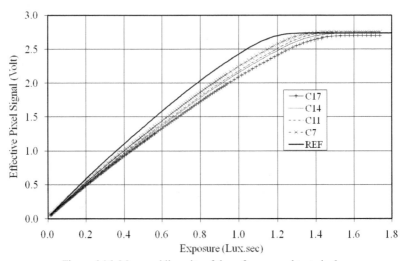

Figure 6.16. Measured linearity of the reference and test pixels.

6.2.6   Photodiode Peripheral Utilization Method (PPUM) Results

Photodiode-type CMOS APS pixels' quantum efficiency was improved by opening number of circular holes on the photodiode diffusion area of a prototype imager. A method called photodiode peripheral utilization method (PPUM) was developed to accommodate pixel performance improvement in a fixed size pixel. Utilizing PPUM, four test pixels with 7, 11, 14, and 17 circular openings, and a reference pixel (REF), were designed, fabricated, and tested in a prototype APS imager made with a 0.5µm, 5V, 2P3M CMOS process. Microphotograph of the prototype CMOS image sensor (PROTO1_1) is shown in Figure 6.17.

Measured pixel characteristics are summarized in Table II. From the test pixels, I was found that PPUM could be used to improve the quantum efficiency and full-well capacity, at the expense of increased dark current and noise level. Compared with the reference pixel (REF), total pixel QE improvement at 390nm was 12 percent for 17 circular openings. Pixel full-well capacity improved 22 percent in the same pixel size, and dark current doubled between reference and C17 pixels.

Table 6.2. Calculated and measured parameters of the pixels with openings and REF1.

Pixel Design	C17	C14	C11	C7	REF	Unit
PD Area	107.5	113.5	119.6	127.6	141.7	μm^2
PD Peripheral	130.1	115.0	99.9	79.8	44.6	μm
Dark Current	6.31	5.29	4.84	4.25	3.15	Ke-/sec.
Conversion Gain	2.80	2.95	3.02	3.19	3.37	$\mu Volt/e-$
Quantum Efficieny	26.22	25.34	25.26	24.58	23.41	% @ 390nm
	51.00	49.41	49.24	47.97	47.44	% @ 550nm
	51.31	50.35	50.72	50.05	48.55	% @ peak
QE improvement	12.0	8.3	7.9	5.0	0.0	$\%QE_{REF}$@390nm
	7.5	4.2	3.8	1.1	0.0	$\%QE_{REF}$@550nm
Sensitivity	2.02	2.09	2.13	2.27	2.44	Volt/Lux.sec
Pixel Full-Well Capacity	621.4	583.1	577.6	545.2	508.7	Ke-
Pixel Full-Well Improvement	22.1	14.6	13.5	7.2	0.0	$\%FW_{REF}$
Pixel Capacitance	57.1	54.2	53.1	50.3	47.5	fF

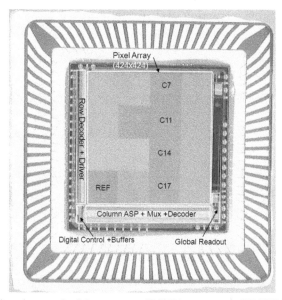

Figure 6.17. Microphotograph of the prototype CMOS image sensor (PROTO1_1) containing reference (REF1) and test pixels with number of openings (C7, C11, C14, C17).

6.3 Photodiode with In-Pixel Capacitance

Pixel photodiode capacitance can be increased by adding a PIP or MIM capacitance in each pixel. However, using MIM or PIP capacitance in pixel is not feasible considering small pixel sizes and layout design rules. In addition, the unit area capacitance of the PIP and MIM capacitances are considerably small to form a large pixel capacitance. An alternative approach is to use MOS type of capacitance to increase the pixel capacity. This approach was adopted in the hybrid three transistor (3T) CMOS APS pixel design with large well capacity [FossumPC]. Imaging capability of the MOS capacitance was utilized to improve overall performance of the hybrid pixel. This new pixel was called hybrid photodiode-photogate (HPDPG) CMOS APS pixel. It was developed to increase pixel full-well capacitance of 3T CMOS APS pixel.

6.3.1   Hybrid Photodiode-Photogate (HPDPG) CMOS APS Pixel

The cross section of the three transistors (3T), hybrid photodiode-photogate (HPDPG) APS pixel is shown in Figure 6.18. Control and readout electronics in the pixel is identical to standard 3T CMOS APS pixel. In HPDPG pixel, NMOS capacitor (NMOSCAP) is surrounded by the photodiode diffusion area which is also the source diffusion of the reset transistor. Gate of the NMOSCAP is connected to a global DC bias voltage. If the bias induced channel potential (Vchannel) of the NMOSCAP was set below the photodiode's reset level (Vreset), two different capacitances (C1, and C2) appears in potential well diagram as depicted in Figure 6.19. The capacitance C1 is associated with the active n+ diffusions of the photodiode while C2 is related to the NMOSCAP. C2 becomes available when the photodiode voltage (Vpd) drops below the channel potential in where the NMOSCAP starts working in the saturation region.

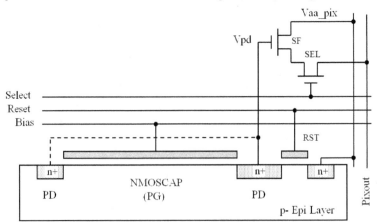

Figure 6.18. Cross section of the hybrid photodiode-photogate (HPDPG) pixel.

164

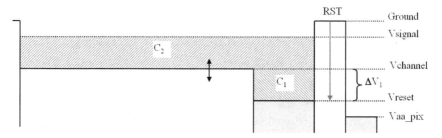

Figure 6.19. Ideal potential-well diagram of hybrid photodiode-photogate (HPDPG) pixel.

Effect of NMOSCAP bias and consequently the channel voltage ($V_{channel}$) on light response characteristics of the HPDPG pixel are depicted in Figure 6.20. If NMOSCAP bias reduced to zero, photodiode pixel saturates at L_{sat1} light level without seeing any extra capacitance. If pixel bias voltage increased, an extra "knee" (1st knee) appears on the response curve. This knee is modulated by the channel potential. Sensitivity of the HPDPG pixel is initially high, and the pixel capacitance is low (C1) until the first knee is reached. After the first knee, second capacitance (C2) is added to the total pixel capacitance decreasing the sensitivity. Second knee is reached when the signal chain saturates. The location of the second knee in terms of light exposure depends on the first knee location and the size of the second capacitance, or the size of the NMOSCAP.

Higher exposure saturation is achieved by increasing the global bias voltage or by increasing the gate area of the NMOSCAP structure during design which results in a desired feature of wide dynamic range pixel operation. Since an NMOSFET structure was used in this design, higher the pixel NMOSCAP bias larger the pixel response range achieved. Opposite control behavior could be used for a PMOSCAP structure.

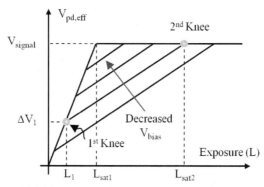

Figure 6.20. Light response characteristics of the HPDPG APS pixel.

165

6.3.2 Operation Principle of HPDPG Pixel

Operation principle of HPDPG pixel is described on simulation setup shown in Figure 6.21. Setup composes of 3T APS pixel transistors (M1, M2, M3), column load transistor (M4), a biased NMOSCAP (PG), column parasitic sampling capacitance (C_S), and a current source (I_{PD}) connected to a photodiode node representing the photo current. A transient simulation was performed with 5.0 volt supply (V_{AA}) and 3.5 volt pixel bias voltages (BIAS). Simulation result of different node voltages along with the normalized pixel capacitance is shown in Figure 6.22.

Pixel photodiode node was reset by activating the RST signal between 0.0 and 0.1 µsec. Pixel photodiode node (V_{PD}) was reset to 3.5 volt. The threshold voltage of the reset transistor (M1) was about 1.5 volt because of the body effect. Same body effect can be observed on the NMOSCAP (PG) transistor which is around 2.0 volt due to the larger device size. As soon as the reset phase ends at 0.1 µsec, charge integration starts. The photo current (I_{PD}) discharges the predefined PD voltage. While V_{PD} drops, the overdrive voltage (V_{BIAS}-V_{PD}) of the NMOSCAP (PG) increases. Thus, the threshold voltage of the NMOSCAP (PG) drops. Right after PD reset, the overdrive voltage is less than the threshold voltage, and the PG works in depletion mode. As a result, overall photodiode capacitance (C_{PD}=C1) is minimal. NMOSCAP starts to go into accumulation mode when the overdrive voltage became closer to the changing threshold voltage. Until this time, photodiode sensitivity is high because of the lower PD capacitance. As soon as strong inversion occurs on the PG, pixel capacitance increases dramatically (Cpd=C2). This causes the pixel sensitivity to drop and the first knee appears on the input-output curve (Vout), as shown in Figure 6.22. Second knee appears when the pixel PD voltage moves below the input range of the pixel source follower.

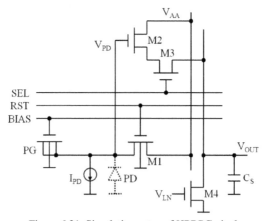

Figure 6.21. Simulation setup of HPDPG pixel.

166

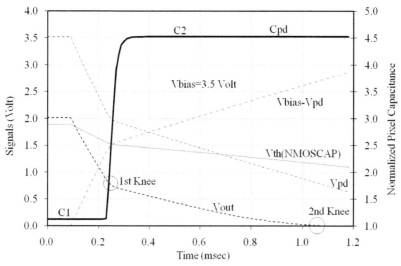

Figure 6.22. Simulation result of HPDPG pixel at 3.5 volt pixel bias.

Photodiode node capacitance increase could be observed on the photon transfer curve. A high conversion gain slope (CG1) is observed until the first knee equivalent effective pixel signal ($V_{PD,eff}$=3.5-2.0~1.5V) was reached. Than a drop in the noise during the capacity transition region, and a lower conversion gain slope (CG2) between first and second knee where the larger well capacity (C2) has been in effect (Vpd,eff=3.5-0.75~2.75V) as depicted in Figure 6.23.

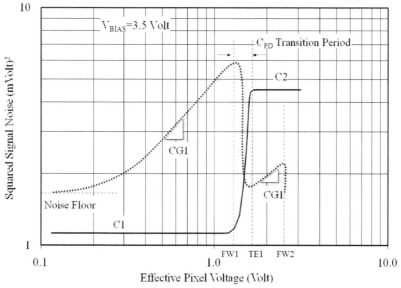

Figure 6.23. Photodiode capacitance transition versus expected photon transfer curve.

Three voltage points on the photon transfer curve can be observed; first well voltage (FW1), capacitance transition end voltage (TE1), and the second well saturation voltage (FW2). The first well voltage (FW1) is the effective photodiode voltage where the capacitance transition from C1 to C2 starts. This transition ends at the transition end voltage (TE1) in where the NMOSCAP overdrive voltage becomes larger than the threshold voltage. At this voltage, the well capacity becomes C2. The second well saturation voltage (FW2) is related to the saturation of the signal chain between pixel photodiode node and the image sensor output. NMOSCAP works in strong saturation regime during this signal range. All these operation regimes and their associated charge capacities are depicted in the Figure 6.24.

Total full-well capacity of the pixel can be calculated by adding the three charge capacities between dark and the second saturation voltage by using the following equation.

$$Q_{pixel} = C_1 \cdot TE1 + C_2 \cdot [FW2 - TE1] + \frac{\alpha}{2} \cdot [C_2 - C_1] \cdot [TE1 - FW1] \qquad [6.5]$$

where

$$TE1 = V(TE1) - Vreset \qquad [6.6]$$

$$FW1 = V(FW1) - Vreset \qquad [6.7]$$

$$FW2 = V(FW2) - Vreset \qquad [6.8]$$

α is related how fast the NMOSCAP changes from depletion to strong inversion. It is typically less that 0.5. Total pixel well depth in number of electrons can be calculated from equation [6.5] by replacing C1 and C2 with the associated conversion gains.

$$N_{pixel} = \frac{\left[C_1 \cdot TE1 + C_2 \cdot [FW2 - TE1] + \frac{\alpha}{2} \cdot [C_2 - C_1] \cdot [TE1 - FW1] \right]}{q} \qquad [6.9]$$

where q is the elementary charge.

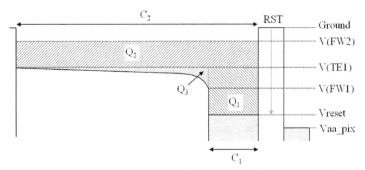

Figure 6.24. Non-ideal pixel capacitance transition and full-well capacity for HPDPG pixel.

6.3.3 Reference Pixel (REF2) Design

A second reference pixel (REF2) was designed for performance comparison. Reference pixel layout and it's A-A* cross section views are shown in Figure 6.25a and in Figure 6.26a. Photodiode region was made up of the n+ diffusion and a donut shaped N-Well regions. N-Well region was placed at the edges of the photodiode to improve the collection efficiency of the photodiode for long wavelength photons. It reduces the overall diffusion related photodiode capacity in both pixel layouts. It is because, the N-well unit area and peripheral capacitances are smaller than that of the n+ diffusion layer for the CMOS process used. Pixel pitch was chosen to be 18μm. It was because; the HPDPG and second reference (REF2) pixels were placed in the same prototype CMOS image sensor chip (PROTO1_1) as the pixels with circular openings and the first reference pixel. Total active photodiode area of the second reference pixel (REF2) was 137 μm^2, resulting in 42% pixel fill factor.

6.3.4 Hybrid Photodiode-Photogate (HPDPG) Pixel Design

Hybrid Photodiode-Photogate (HPDPG) pixel layout and it's A-A* cross section views are shown in Figure 6.25b and in Figure 6.26b. All imaging structures including the NMOSCAP poly-silicon gate were drawn circular. Pixel electronics in HPDPG pixel are same as the second reference pixel. Bias line of HPDPG pixel was drawn on first metal layer horizontally throughout pixel array. In HPDPG pixel layout, circular NMOSCAP area occupies 58 μm^2 of total pixel area. Thus, pixel fill factor drops to 24%. Although, NMOSCAP contributes photon collection, it was expected to be minimal.

Microphotograph of the prototype CMOS imager is shown in the Figure 6.27.

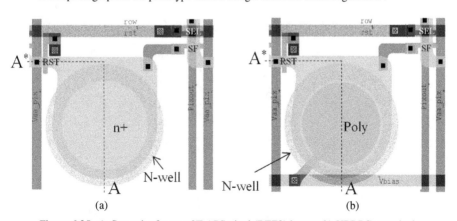

(a)   (b)

Figure 6.25. a). Second reference 3T APS pixel (REF2) layout, b) HPDPG test pixel.

169

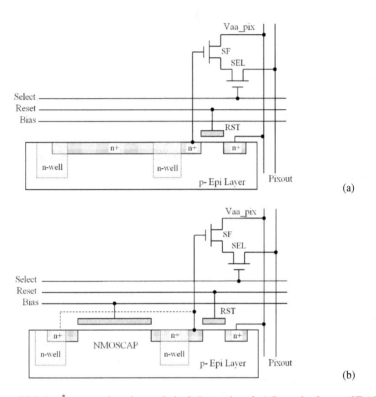

(a)

(b)

Figure 6.26. A-A* cross section view and pixel electronics of; a) Second reference 3T APS pixel (REF2) b) HPDPG test pixel.

Figure 6.27. Microphotograph of the prototype CMOS image sensor (PROTO1_1) containing reference (REF2) and HPDPG test pixel.

170

6.3.5   Measurement Results

6.3.5.1   *Conversion Gain and Full-Well Capacity*

Conversion gain of the second reference and the HPDPG pixels were measured by using the photon transfer curve method. Measured conversion gain (CG) of the second reference pixel (REF2) was $4.85 \mu Volt/e^{-}$. This resulted in 372,650 electrons full-well capacity and 33.8fF photodiode capacitance on 1.76 volt pixel voltage swing.

As expected, HPDPG pixel photodiode capacitance depends on the NMOSCAP bias voltage. Measured photon transfer curves of the HPDPG pixel for different NMOSCAP bias voltages are shown in Figure 6.28. Pixel capacitance transitions or knee points are clearly visible for high bias voltages in the photon transfer curves. Pixel capacitance values, C1 and C2, could be measured indirectly by measuring the conversion gain in their dedicated operation regions. Measuring C1 values from photon transfer curves were relatively easy. However, because NMOSCAP gate capacitance in saturation region was very large, C2 measurement was quite difficult except for pixel bias values between 4 and 5 volts.

Extracted conversion gains of the HPDPG pixel for bias voltages between 0.1 and 5 volts are shown in Figure 6.29. Two conversion gain factors were measured. First one is between dark and first knee start points. The second one is between transition end and second knee start points. The first one is associated with the pixel capacitance C1, and the second one is associated with C2. Capacitance C2 was measured only at high bias voltages. For each bias level, FW1, TE1, and FW2

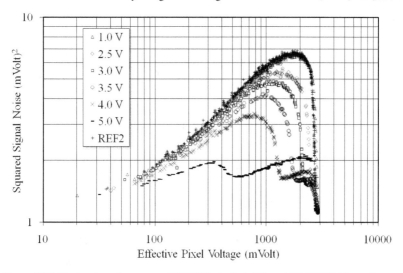

Figure 6.28. Photon transfer curves of HPDPG pixel at different MOSCAP bias voltages.

171

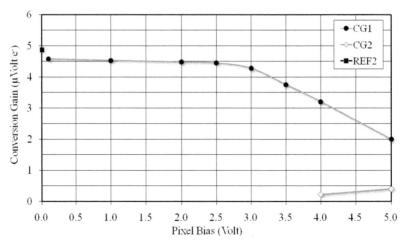

Figure 6.29. Measured conversion gain of HPDPG and REF2 pixels at different NMOSCAP bias voltages.

voltages were determined to calculate C1 and C2 related full-well capacity using equation [6.9]. Conversion gain associated with the C1 could be adjusted between 4.58μVolt/e- and 2.00μVolt/e- for bias voltages between 0.1 volt and 5.0 volt, respectively. Conversion gain for C2 was measured at 4.0 and 5.0 volt bias voltages and they were 0.23μVolt/e- and 0.4 μVolt/e-, respectively.

Extracted pixel full-well)capacities associated with the C1, and C2 are shown in Figure 6.30. For increased bias voltage, C1 contribution to the total well depth reduces, because of the reduced first knee start point, FW1. Although, the conversion gain of the C1 drops with increased

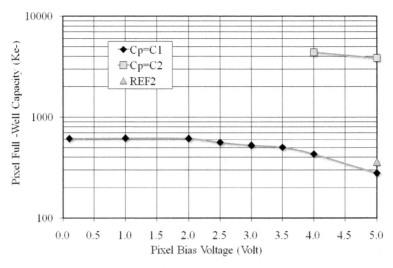

Figure 6.30. HPDPG Pixel well depths associated with C1, and C2.

172

bias voltage, it could not compensate the fast dropping of the first knee location. The transition end location (TE1) moves lover for increased pixel bias voltages. This effectively increases the voltage difference (FW2-TE1) that the second capacitance operates (C2) until the second knee reached.This increase is compensated by the increase in the C2. Thus, C2 related full-well capacity component decreases with increased bias voltages. Overall, HPDPG pixel structure provides an order full-well capacity increase in standard 3T CMOS APS pixels. At 5volt bias, 4 million electrons pixel full-well capacity was achieved.

6.3.5.2   Quantum Efficiency (QE)

Quantum efficiency (QE) of the second reference (REF2) and the HPDPG pixels were measured, and the results are shown in Figure 6.31 and Figure 6.32. QE of the HPDPG pixel was measured at 0, 3, and 5-volt pixel bias voltages. About 6.5% peak to peak quantum efficiency variation was observed between different bias voltages for HPDPG pixel. Measured average peek quantum efficiency for HPDPG and second reference pixels were 29.5% and 49.6% at 500nm, respectively. Average quantum efficiency loss between the second reference (REF2) pixel and the HPDPG pixel between 390 nm and 700 nm are shown in Figure 6.31. Average 33.7% QE was lost in HPDPG pixel comparing with reference pixel between 390 nm and 700 nm. This drop was expected because of the opacity of NMOSCAP gate poly. NMOSCAP reduces pixel fill-factor fron 42% to 24%, a 40% reduction. Thus expected QE reduction was 40% as oppose to the measurement of 34%. Thus the NMOSCAP contribution to pixel QE is about 6%.

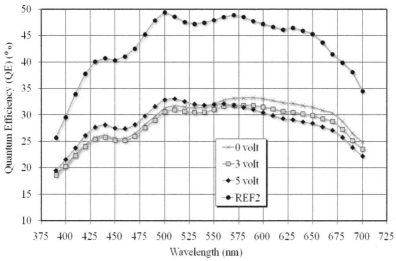

Figure 6.31. Quantum efficiency measurement results of the second reference (REF2) and the HPDPG pixels.

173

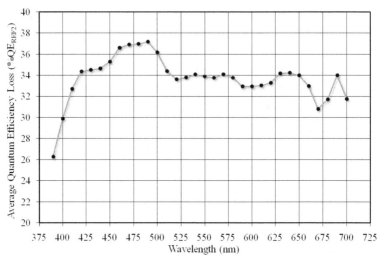

Figure 6.32. Average quantum efficiency loss measurement results of the HPDPG pixel.

6.3.5.3   Dark Current

Dark current voltage and associated dark electrons were measured, and the results are shown in Figure 6.33 for different pixel bias voltages for the HPDPG pixel. The dark images at all bias levels see the pixel capacitance C1 in the HPDPG pixel. Thus, only the conversion gain associated with the C1 was used to convert the dark current voltages into number of dark electrons. Dark current voltage and electrons the second reference pixel was measured first. It showed 13.43 mVolt/sec. dark current voltage, and 2770 e-/sec dark electrons. As seen in Figure 6.33, dark current electrons for the HPDPG pixel drop until the pixel bias voltages around 3.5 volt and increases after that. Comparing with the reference pixel, dark current electrons of the HPDPG pixel increases from 5950 to 6350 e-/sec. for pixel biases between 0.1 and 5.0 volts, respectively. This is an increase between 2.15 and 2.30 times of the second reference dark electrons. Comparing with the dark current voltages, the HPDPG pixels showed smaller dark current voltage level at 5 volt pixel bias condition. Measured dark current voltage at this bias condition was 12.7 mVolt/sec. This drop was because of the smaller photodiode region that make up off the C1 capacitance. Although the photodiode area was dropped 45% in the HPDPG pixel, the dark current generation centers in the photodiode edges (peripherals) was the same as the second reference pixel. The dark electron increase could also be attributed to the unbalanced drop on the dark current voltage and the conversion gain of the HPDPG pixel up to the first knee point. This is because of the bias dependent capacitance contribution of the NMOSCAP (eventhough it is OFF) to the photodiode region and the C1.

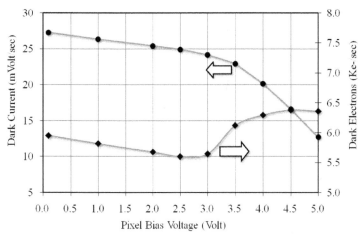

Figure 6.33. Dark current in volt and in electron measurement results for pixel HPDPG.

6.3.5.4 Linearity and Sensitivity

Sensitivity of HPDPG pixel was measured and results are shown in Figure 6.34 for pixel biases of 0.1, 3.0, 4.0, 4.5 and 5.0 volts. Bias depended range expansion is clearly seen from the measurement. As expected, first and second knee locations are also changed with changing bias conditions. Sensitivity of HPDPG pixel before the first knee drops fast for increased pixel bias voltages unlike sensitivity between first and second knee as shown in Figure 6.35. Sensitivity drop before the first knee is associated with the drop in the photodiode capacitance, C1, for increased pixel bias level as observed in the photon transfer curve of the HPDPG pixel. C1 related sensitivity

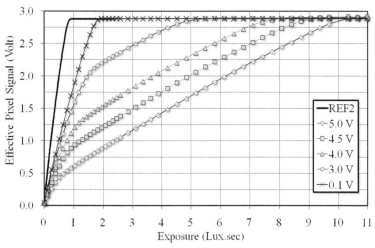

Figure 6.34. Sensitivity measurement results of the HPDPG and reference pixels.

175

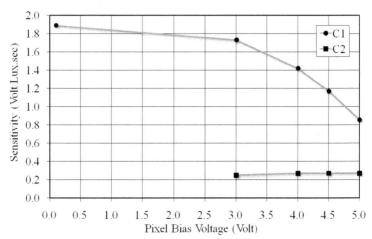

Figure 6.35. Sensitivity variation of the HPDPG pixel before (C1) and after (C2) the first knee on the sensitivity plot.

changes between 1.9 and 0.8 Volt/Lux.sec, while the C2 related sensitivity changes very little between 0.25 and 0.27 Volt/Lux.sec. Second reference (REF2) pixel sensitivity was 3.75 Volt/Lux.sec. At 0 volt and 5 volt bias levels, HPDPG pixel shown two to five times drop in sensitivity, respectively. This is)because of the increased pixel photodiode capacitance and reduced photodiode pixel fill factor about 40% in the HPDPG pixel.

Fill factor of the photodiode region could be increased by placing the NMOSCAP device to somewhere in the pixel that may be otherwise unused such as under the routing metal lines. This way both sensitivity and the quantum efficiency could be improved.

6.3.6 APS Pixel with In-Pixel Capacitance Discussion

A CMOS pixel called hybrid photodiode-photogate (HPDPG) APS pixel was developed to improve the pixel full-well capacity and dynamic range. HPDPG pixel composes of n+ photodiode diffusion area and a biased NMOSCAP in each pixel. NMOSCAP was biased such a way that there are two knee appears in the light response curve. Locations of the knee points were adjustable with the pixel bias voltage that was connected to the gate of the NMOSCAP device. When the NMOSCAP was biased at strong inversion region by setting the bias voltage to 5 volt, more than 4 million electrons (or 10x improvement over reference pixel's full-well) full-well capacity was achieved at saturation while light response range was extended about 5 times providing wide-dynamic range pixel operation.

6.4 Summary

Two main investigations were completed to increase the both pixel capacity and quantum efficiency of 3T CMOS APS pixel. Pixel capacity, or well depth, increase was achieved by two methods; n+ active photodiode peripheral utilization for certain feature size CMOS processes, and by including an NMOSCAP in-the pixel that is connected in parallel with the n+ active photodiode region.

Peripheral utilization and capacity increase was achieved by opening a number of circular openings on the photodiode n+ active area. These opening sizes were chosen such a way that the depletion overlap induced peripheral capacity drop was avoided. To test this idea, four test pixels with 7, 11, 14, and 17 circular openings, and a reference pixel (REF1) were designed, fabricated, and tested. A 0.5 μm, 5 volt, 2P3M CMOS process was chosen for the test chips avoiding CMOS processes that provide very high unit active area junction capacitances and very low peripheral unit capacitances. From the test pixels, it was found that the photodiode peripheral utilization method (PPUM) could be used for improving the pixel full-well capacity with the expense of increased dark current. Peripheral utilization method also provides better blue spectral response trend. Compared with the reference pixel (REF1), total pixel well depth increase was about 25% for 17 circular openings. QE improvement at 390 nm was more than 10%.

A new pixel type called hybrid photodiode-photogate (HPDPG) APS pixel was developed to improve the pixel well depth dramatically. HPDPG pixel composes of n+ photodiode diffusion area and a biased NMOSCAP in each pixel. NMOSCAP was biased such a way that there are two knee appears in the light response curve of the new pixel. Locations of the knee points were adjustable with the pixel bias voltage that was connected to the gate of the NMOSCAP device. When the NMOSCAP was biased at strong inversion region by setting the bias voltage to 5 volt, more than 4 million electrons well depth was achieved at pixel saturation for the HPDPG pixel. Two times more dark current electrons were observed in the HPDPG pixel. The HPDPG pixel presented tree major shortcomings. One was the reduction of the quantum efficiency due to the fill factor loss by the large NMOSCAP gate area. The other one was the increase in dark current. Third one was the reduced sensitivity because of the fill factor reduction.

CHAPTER 7 CMOS IMAGE SENSOR DESIGN METHODOLOGIES

Large format, CMOS APS image sensor design issues and development methodologies were addressed in this chapter. Large area image sensors design limits were investigated. Design methodologies were reviewed and a new methodologies were developed to address large format image sensor design issues for so-called CMOS processes with stitching option.

7.1 CMOS Image Sensor Design Methodology (CIS-DM)

CMOS image sensor design process could be divided into interdependent tasks that have to be addressed methodically as depicted in Figure 7.1. They are market research, technology evaluation for foundry and process selection, pixel design, image sensor electronics design, fabrication, testing, and system integration. The foundry/process selection might have multiple repetitions for different foundry services if the methodology is used by a fables image sensor company. Even, it is used by a company which has its own fabrication capability, selecting process node among different ones is necessary. In this case the cost of final product and marketing research plays an important role for profitability of the product.

Although, imaging pixel is one of the major sub-blocks of an image sensor integrated circuit (IC), we considered its design methodology separately from the design methodology of image sensor electronics. It was because, pixel design encompasses not only electrical but also optical and

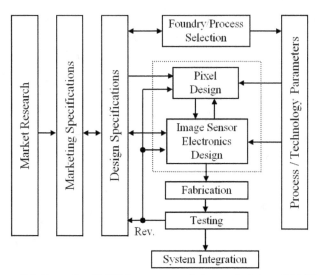

Figure 7.1. Generalize CMOS image sensor design methodology (CIS-DM)

179

process level design and simulations unlike the other blocks of image sensor application specific integrated circuits (ASICs) such as analog-to-digital converter (ADC), digital-to-analog converter (DAC), analog and digital signal processing circuits, etc.. Sometimes, especially for companies that have its own fabrication capability, might develop pixel array and deliver for integration independent from the imager electronics with dedicated process and device futures, such as a mask defining pinned-photodiode doping level or an NMOS device which has different threshold voltage than that of used in analog or digital part of the imager electronics.

CMOS image sensor design starts with market research. If market research results in feasibility and profitability of the proposed imager, than image sensor specifications are identified. Most of the times marketing specifications changes depending on the competition or market conditions. Design has to negotiate with marketing to generate reasonable design specifications that could be achieved with current know-how in imaging electronics and pixels of the company. During this time, foundry (through which image sensor would be manufactured) and process selection process starts. Foundry and process selection process requires a technology evaluation methodology to map process end technology parameters to image sensor design requirements. After design specifications were fixed, and a process and a foundry were chosen, pixel and image sensor designs would be started.

Physical, electrical, and optical characteristics of pixel layouts are estimated based on evaluated foundry's process parameters through process and electro-optical simulations and modeling. Since pixel design requires multi disciplinary cooperation and coordination, its design has to be dealt with through a pixel design methodology. At the same time, design of the image sensor functional blocks progresses based on pixel design, image sensor design specifications and technology parameters of the selected process. A standard CMOS mixed-signal design methodology could be used in image sensor electronics design. After the image sensor IC is designed and fabricated, electrical and optical characteristics are measured. Based on these measurement results either a new revision is planned or the design is moved into mass production after customer evaluation. Revisions could be done on design specification, or on the imaging pixel, or only on a small part of the sensor circuits. Characterization results could be used for developing technology evaluation and pixel parameter estimation models.

7.2 Technology Evaluation Methodology

A technology evaluation methodology is depicted in Figure 7.2. Main purpose of this methodology is to select most appropriate foundry services and process technology for the image sensor under development. This is the first and the most important step for a fabless image sensor design company during an image sensor design.

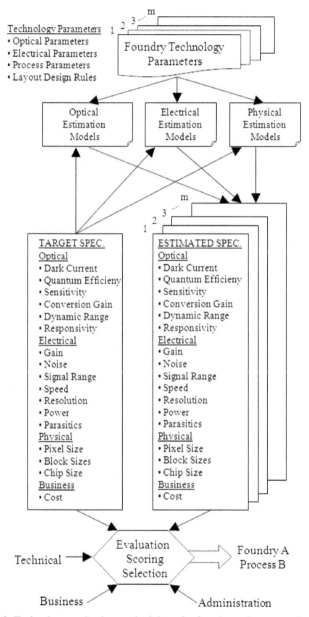

Figure 7.2. Technology evaluation methodology for foundry and process selection.

Technology evaluation methodology starts with building databases for process related technology parameters of foundry services that may or may not provide CMOS image sensor (CIS) oriented processes. Optical, electrical, and process related parameters and models as well as physical layout design rules of evaluated processes has to be collected in these databases. They are used along with target specifications of proposed image sensor to estimate optical, electrical, and physical properties of candidate pixel technologies. Modeling and simulation tools are used to determine expected pixel performances for each process technology.

A most appropriate foundry and process technology is selected with the help of technical, administrative, and business organizations of the company and final design specifications are updated for detailed electronics and pixel design. Consequently, short and long-term design schedules are built for successful completion and launch of the image sensor product.

Technology evaluation process sometimes is a very cumbersome process considering all the requirements of a high-quality image sensor. Because none of the standard CMOS foundries could be able to satisfy all the optical and electronic requirements of a CIS ASIC. They might provide processes with multi or very low threshold devices with very low supply voltage for low power mixed-signal ASIC design, but may not provide low junction leakage current for low dark current pixel operation. They might provide four layers of metal routing for dense layout design, but not the opacity of layers that an impinging photons to pass to reach the photo collection sites. Rarely, but sometimes, they provide very good electrical and optical performance, but very poor manufacturing yield, resulting in non-profitable products. Thus, it is not an easy process to select a CMOS process technology for a successful image sensor design, especially for fabless image sensor design companies. Of course, this approach assumes that optical and process parameters of the evaluated or selected processes are available to users from foundry services. Because, most of the times, doping levels of the active areas, diffraction indexes of the dielectric layers, etc. carries proprietary process information, and considered proprietary information by manufacturers.

Optical, electrical, and physical estimation models are essential part of image sensor performance prediction before design starts. These models could be generated through test pixels development and characterization in selected process technologies, or through electro-optical simulation tools. For example, if proprietary process fabrication steps and doping profiles of the process are known, it is very easy to estimate the quantum efficiency, dark current, full-well capacity and signal range of pixel using process simulation tools. If they are not provided by the fabrication service or company, a dedicated test run is necessary for both electro-optical and electronic modeling. In some applications, technology evaluation may result in specialized CMOS processes or techniques such as stitching, backside thinning, phosphor coating, etc., for companies that have its own fabrication capabilities.

7.3 Pixel Design Methodology

Imaging pixel is the most important element in an image sensor ASIC. Thus, it could not be designed as a straightforward circuit block. Pixel development effort could be modeled by two strategies. One is in-house pixel development strategy. The other one is fabless pixel development strategy.

In in-house pixel development strategy, both pixel and manufacturing process were designed and optimized for electro-optical performances in company owned facilities. This development strategy requires lots of expertise, effort, time, and funding. Almost all of the CCD manufacturers follow this development strategy. It is mainly because CCD quality is directly related to the quality of process and control of process parameters. During past couple of years, in-house development strategy is also adapted by CMOS image sensor companies. This is mainly because of the increased market share of CMOS imagers in the global image sensor market. Thus, in-house development becomes feasible and profitable for CMOS image sensor in high volume, high margin applications.

In fabless pixel development strategy, image sensor design house uses a CMOS foundry services. This foundry provides high performance CMOS processes optimized for either digital or mixed-signal IC manufacturing, but not necessarily for an image sensor IC manufacturing. Since, most of the time, processes have propriety manufacturing steps and features, foundry users could not have deep knowledge and control of the process they selected to use. They might only be provided with the process design rules and the models to design their circuits during the design process. Thus, the fabless image sensor design companies spent most of their design cycles for optimizing and tuning their imaging pixels. To do so, they generally produce multiple engineering runs to model the gap between process that is available and the image sensor quality metrics such as dark current, sensitivity, quantum efficiency, etc. Vehicle for this modeling is to use an image sensor architecture that contains basic signal processing circuits and holds multiple pixel designs. So-called, Fully-Flexible Open Architecture (FFOA) that is introduced in chapter 5 is one of the most used imager architecture for this purpose.

Current trend in CMOS foundries is to provide better mixed-signal CMOS processes or CMOS image sensor (CIS) processes that have optimized electro-optical properties for solid-state imaging. CIS foundries provide not only the electrical properties of the active devices but also the optical properties. In some cases, imaging pixel itself is provided for better image quality and manufacturing yield. This reduces workload of image sensor design companies during development phase. In this study, only fabless pixel development methodology (FPDM) was explored and used during prototype CMOS image sensor design.

7.4 Fabless Pixel Design Methodology (FPDM)

Proposed fabless pixel design methodology (FPDM) composes of seven sub tasks or processes as depicted in Figure 7.3. These are model development, pixel performance estimation, pixel parameter and type extractor, pixel electronic design, pixel layout design, pixel parametric extraction, and electronic and photonic simulation.

Most of the application specific image sensors dictate pixel type, size and other parameters such as dynamic range, spectral response, sensitivity, conversion gain, etc. If it is not specified in design specifications, pixel type and size are determined first. It is done by extrapolating pixel performances for given pixel type and size based on the foundry provided or design-house build pixel and process models. After pixel type, size, and other physical, electrical, and optical design parameters determined, electronic and physical design of the pixel starts. In pixel electronics design, basic active devices such as switch and amplifier transistors are designed, and preliminary

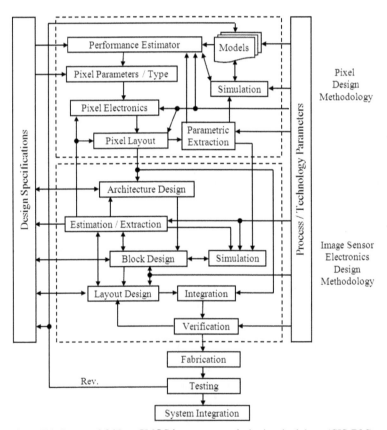

Figure 7.3. Proposed fabless CMOS image sensor design methodology (CIS-DM).

184

simulations are completed. After the physical layout design, pixel circuit parasitics and physical structures are extracted with the provided technology information for electrical, optical and process simulations. Fine tuning of the pixel parameters was done by multiple iterations in the pixel performance estimation, layout design and simulation paths.

Pixel design methodology has two-way relations with the image sensor electronics design methodology as depicted in Figure 7.3. Especially pixel pitch plays an important role in image sensor's signal chain architecture. Sometimes, feasibility or requirements of some blocks in the image sensor electronics affect pixel electronics or pixel layout. For example, pixel source follower device sizes determine pixel source-followers (PSF) gain, linear I/O range and noise performance in CMOS APS image sensors. Based on the frame rate, supply voltage, analog signal chain (ASC) architecture, and dynamic range requirements, PSF device sizes has to be arranged such a way that the 1/f noise component is reduced and the ASC gain is maximized with wide linear operation range.

7.5 Image Sensor Electronics Design Methodology

Image sensor electronics design methodology in not very different from that of a top-down application specific integrated circuit (ASIC) design methodology. Difference is that architecture and the functional block definitions and specifications of the imager electronics are not only defined and affected by design requirements but also by pixel design as shown in Figure 7.3.

Image sensor electronics design starts with architecture design and definition. Imager architecture design depends on design specifications and pixel design. Leading design parameters are frame rate, power consumption, die size, cost, pixel type, pixel pitch, pixel array orientation, and output format of the image sensor. Image sensor architecture defines analog and digital processing circuit blocks and their functionalities between pixel photosensitive site and image sensor's outputs.

Analog signal processing circuits between pixel photosensitive site and chip outputs is called analog signal chain (ASC). Typical block diagram of a series ASC is shown in Figure 7.4. Pixel composes of photo conversion site (PD) and part of the pixel source follower amplifier. An n-type source follower is typically used in CMOS APS imagers as pixel amplifier and called pixel source follower (PSF). PSF buffers photodiode signals to the column sample and hold (CSH) circuitry. Depending on the signal chain design, CSH signals pass through a column amplifier and column select switch to reach a global programmable gain amplifier (PGA) circuits. Most of the analog processing such as programmable gain, offset correction, level shifting, correlated double sampling (CDS), and signal buffering are done before the column signals were sent to analog-to-digital converter(s) (ADC).

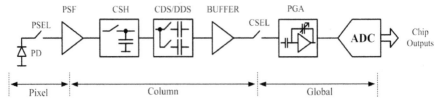

Figure 7.4. Typical analog signal chain (ASC) block diagram.

ASC block boundaries and location of analog signal processing operations depend on the image sensor architecture design. Three major types of ASC architectures are used in CMOS APS imagers; serial, column-parallel, and pixel-parallel as discussed in section 5.3.

After the architecture definition, block level design specifications are generated. During block design, each block is simulated stand alone and/or in multiple block formations. After the block requirements meet, layout of each block is designed. Fine-tuning of the block performances stand alone and in defined architecture is done by extracting and simulating the parasitics of the block layouts. After the parasitic simulations are concurred with the block and system requirements, block layouts are integrated and the layouts are verified against the schematics and the process requirements. After the full chip verification, the full chip simulation is performed and the chip is taped out to foundry for fabrication.

After the fabrication, the imager chip is packaged and tested. Optical and electrical characteristics of the image sensor are measured and verified during testing. If imager meets all of the design specifications, the imager would be sent to customers for evaluation as engineering samples. Otherwise, a new revision is planned while the engineering samples of the current version are shipped to customer for evaluation with the promise of fixed next revision of the imager. After customer commitment the design go into mass production and system integration.

7.6 A Special CMOS Fabrication Process: Stitching

Photolithography methods are used for patterning devices and interconnect layers onto a silicon wafer during CMOS integrated circuit (IC) fabrication. They are formed on the wafer by using mask exposures, repeated reticle exposures, or repeated reticle sub-section exposures methods as shown in Figure 7.5.

Mask exposure method is typically used in early CMOS fabrication processes that have 1.0 μm or larger device sizes. A wafer size mask is used for patterning. Full wafer patterning takes place at one time without any optical focus. In this method, the maximum individual chip size, or the die size is limited only by the size of the mask or the wafer. This patterning and fabrication mask is shown in Figure 7.5a.

186

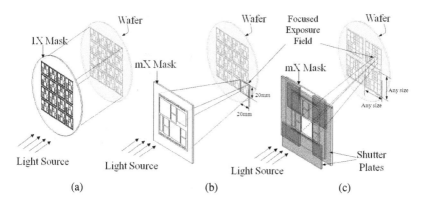

Figure 7.5. Circuit pattern formation on silicon wafer through a) mask exposure, b) repeated reticle exposure, c) repeated reticle sub-section (stitching) exposure.

In today's sub-micron CMOS processes, on the other hand, modern stepper photolithography equipments are used to form circuit and interconnect patters on a wafer by focusing and repeating a much smaller exposure field or reticle. A typical reticle field is about 20 mm by 20 mm. An M times (typically, 5x for 6-inch wafers) focusing mechanism is used on reticle field to repeat the patters. In this method, the maximum individual chip die size is limited by the reticle size of 20mm by 20mm as shown in Figure 7.5b.

Stitching method overcomes the stepper limitations and allows die sizes larger than the reticle exposure field [Kreider95, McHung00]. It allows several design structures to be physically merged on the same wafer in order to make a single large die. The stepper exposes one stitching circuit block at a time by the precision alignment method with slight overlaps. Thus, the blocks that are going to be repeated have to be designed in two, three, or four side stitchable fashion to allow a formation of large die ICs. In stitching processes as big as a wafer size die could be manufactured as a single chip as shown in Figure 7.5c.

7.7 Stitching Design Methodology

A new design methodology has to be utilized for achieving first silicon success for special CMOS processes such as CMOS process with stitching option. It is because, the blocks that are going to be repeated in stitching process have to be designed in multiple side stitchable fashion with strict block sizes and functionalities to allow a formation of large IC die. Proposed methodology is shown in Figure 7.6. Stitching design methodology is similar to the CMOS image sensor design methodology described in the previous sections. Only difference is the added stitching pixel and block design processes. They add more complexity on design process. Because they affect the decisions made in both pixel design and image sensor electronics design processes.

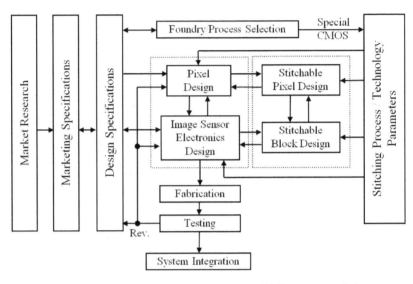

Figure 7.6. Stitching design methodology for CMOS image sensor design.

After technology evaluation process was completed and a special CMOS process and foundry was selected, image sensor design process starts. What makes the special or stitching CMOS process to be selected is the die size that was estimated based on the design specification such as pixel array size and the pixel pitch during the technology evaluation process.

Design mainly revolves around the stitching design rules first. They are used for determining the stitching block and pixel dimensions. Typically, a sub-array of pixels has to be repeated more than the other stitching blocks. Thus, a four side stitchable pixel layout has to be designed. Other block layouts could be designed with two side stitching features. Third side could be stitched by using small arrays of four side stitchable pixels. Again, design of the pixel and its layout play a central role in stitching design methodology for determining flexibility and quality of the image sensor under development. After stitchable pixel size was determined based on stitching design rules, pixel design is completed by using the pixel design methodology described before. Image sensor electronics is designed by following the methodology described in section 7.2 with the added complexity of stitching block sizes and functionalities. Imager chip architecture is defined and partitioned into functional sub-blocks. Then, based on the stitching block dimensions, each functional block locations in the stitching blocks are identified. Then, each functional block is designed. Stitching design rules are used during the chip integration and verification process to make sure that a seamless stitching block transitions are occurs during fabrication process.

After blocks were designed, stitching reticles are generated. These reticles could be used to produce any array size of the same designed image sensor. This leads reduction in manufacturing cost; and adds flexibility to the product mix of a design house. Because, with same mask set, product matrix becomes a linear fractions of the sub-pixel-array block sizes. For example, design house could provide exactly the same image sensor with $mK \times nK$ formats. K is the sub-pixel-array block size that was designed with four side stitching features, and m and n are the fractions that a customer requests.

7.8 Summary

General CMOS integrated circuit (IC) design methodologies, CMOS image sensor design methodologies for fables companies were reviewed in this chapter. A relatively new IC fabrication techniques, the stitching, was introduced. New CMOS pixel and imager design methodologies were proposed. These methodologies were used both designing standard CMOS image sensor ICs and specialized CMOS image sensors with stitching option. Stitching design methodology is put to work for designing very large area image senor. Details of this effort are presented in next chapter.

CHAPTER 8 LARGE FORMAT CMOS APS IMAGER DESIGN: A CASE STUDY

New stitching image sensor design methodology was introduced in chapter 7 and put to work with a case study in next this chapter. A CMOS APS image sensor with nearly 16.85 million pixels (4096 x 4114) and more than 1 million electrons (1Me-) pixel full-well capacity was designed, fabricated, and characterized. Each of the design steps from design specifications to chip characterization was explained under the new design methodology. Feasibility of designing and fabricating a CMOS APS image sensor of 76.08mm x 77.55mm die size in a single 6-inch wafer was demonstrated by using the stitching design methodology. Measurement results of the 4Kx4K CMOS APS image sensor were presented.

8.1 Foundry and Process Selection

First, a foundry and process were chosen based on the initial specification of the imager array size, full-well capacity, dynamic range, cost, availability of process, and schedule. The model equations [4.65] and [4.18] were used to determine the process minimum feature size and pixel pitch for given foundry service and processes. As seen in Figure 4.40, minimum pixel pitch of at least 10 μm is required for 1Me- full-well capacity for any process feature sizes. As a result, it was found that with these initial specifications, none of the standard CMOS processes with 20mm by 20mm reticle size could be used to build the specified image sensor. Thus, a special 0.5 μm, two poly three metals (2P3M), 5 volt CMOS CIS process with stitching process option was chosen. Design rules, yield profiles, and the technology parameters were obtained from the foundry service for designing the imager.

8.2 Pixel Size and Sub-Array Size Design

Based on the technology parameters of the selected process, at least a 17.8 μm pixel pitch is required to achieve 1Me- pixel well depth on 1-volt pixel signal swing. Based on the stitching layout design rules, on the other hand, pixel pitch was found to be between 17 and 19 μm. This dictated the maximum size of the repeatable pixel block in the stitching process. Thus, the pixel pitch was selected to be 18 μm for the design. This selection resulted in a repeatable 512 by 512 pixel sub-array. It was stitched or repeated 64 times to form a 4096 by 4096 (4Kx4K) CMOS APS image sensor array. Pixels are used as border element for surrounding stitchable blocks. These pixels were also contains dark pixels to signal processing. Border pixels were chosen to be imaging ones for correct stitching operation during fabrication. As a result, total pixel array becomes larger than the targeted 4096 x 4096. Total dark and open pixels were 4112 x 4114 of which only 4096 x 4114 could be used for scene imaging.

8.3 Stitchable HPDPG Pixel Design

A new version of the HPDPG pixel was designed for 4Kx4K image sensor based on the measurements done on the prototype test pixels. NMOSCAP gate area was determined by plotting the measured NMOSCAP gate area in the test pixels versus the measured conversion gain and full-well capacity for 1 million electrons as shown in Figure 8.1. Required NMOSCAP gate area was found to be 50 μm^2 for 1Me- well depth on 1 volt pixel signal swing.

Area and peripheral of the n+ photodiode region of the HPDPG pixel were determined by evaluating the measurement results of the prototype pixels with 7, 11, 14, and 17 circular openings and the first reference pixel. It was found that increasing the peripheral area while decreasing the total active area increases both visible and blue spectral quantum efficiencies as shown in Figure 8.2 and Figure 8.3, respectively. Increased peripheral length reduces the area to peripheral ratio for fixed pixel size and increase the dark current as shown in Figure 8.4. Quantum efficiency at 390 nm and the dark current were modeled based on the test pixel measurement results with respect to number of 1.6 μm^2 circular openings or area to peripheral ratio. Extrapolated number of opening versus quantum efficiency in 390 nm and 550 nm, and the dark current and well depth are shown in Figure 8.5 and Figure 8.6, respectively. 12 openings were placed on photodiode area to increase the quantum efficiency of the pixel to compensate the area sacrificed for the NMOSCAP.

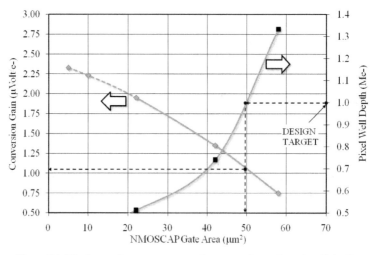

Figure 8.1. Pixel capacitance area versus the conversion gain and well depth.

192

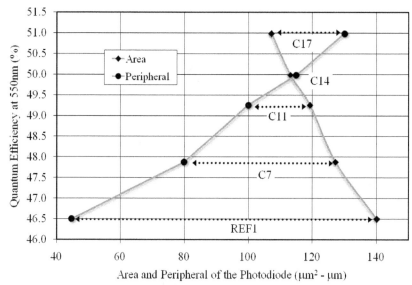

Figure 8.2. Pixel area and peripheral versus quantum efficiency at 550 nm.

Expected quantum efficiency of the pixel was 45.6 % at 550nm, and 23.7% at 390 nm. Estimated dark current electrons were 24,000 electrons per second. Well depth was estimated to be around 1.05 Me-. Based on these numbers, layout of the pixel was designed with four side stitching capability as shown in Figure 8.7.

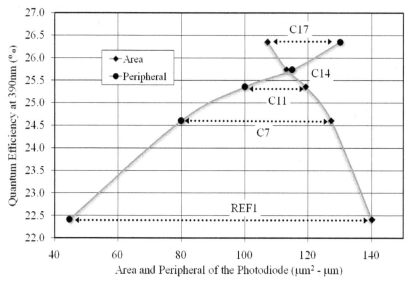

Figure 8.3. Pixel area and peripheral versus quantum efficiency at 390 nm.

193

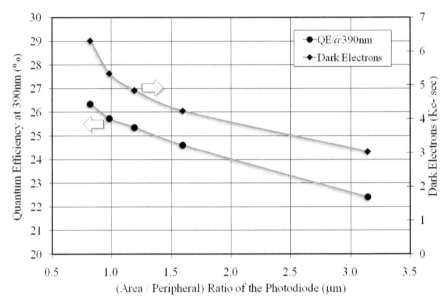

Figure 8.4. Test pixel measurement results of quantum efficiency (at 390 nm) and dark current electrons versus area to peripheral ratio of the photodiode region.

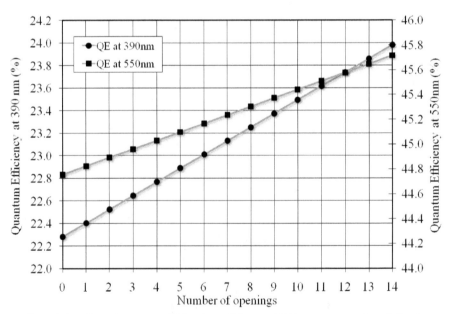

Figure 8.5. Estimated quantum efficiencies at 390nm and 550nm versus number of circular openings for designed pixel.

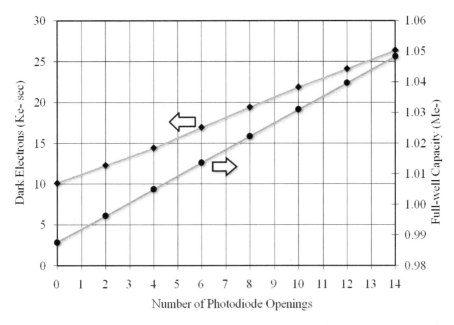

Figure 8.6. Estimated dark current and well capacity of the designed pixel versus the number of circular openings.

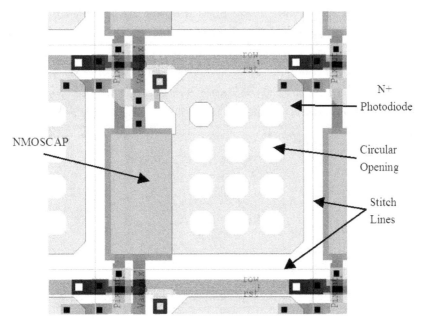

Figure 8.7. Layout of the four side stitchable HPDPG pixel (18 μm x 18 μm).

8.4 Stitching Block Dimensions

After the pixel size was determined, each stitching-block sizes and functional details were determined based on stitching design rules, technology parameters, and design specifications. Block names and number of repetitions of each block on the imager die is shown in Figure 8.8.

A	B	B	B	B	B	B	B	B	C
D	E	E	E	E	E	E	E	E	F
D	E	E	E	E	E	E	E	E	F
D	E	E	E	E	E	E	E	E	F
D	E	E	E	E	E	E	E	E	F
D	E	E	E	E	E	E	E	E	F
D	E	E	E	E	E	E	E	E	F
D	E	E	E	E	E	E	E	E	F
D	E	E	E	E	E	E	E	E	F
G	H	H	H	H	H	H	H	H	I

Figure 8.8. Block names and number of repetitions on the die.

Design composes of 9 blocks that was named A, B, C, D, E, F, G, H, and I. Blocks D and F repeated 8 times on y-direction. Blocks B and H repeated 8 times on x-direction. Block E repeated 64 times; 8 times on both x-direction and y-direction. Blocks A, C, G, and I repeated only once at the corners of the chip die. Block E contains the sub-pixel-array with 512 by 512 array format. Based on the stitching block design rules provided by the foundry, each block sizes were determined as shown in Figure 8.9. Based on the pixel pitch of 18 μm and other design parameters, total chip die size was found to be around 76.08mm x 77.55mm. This resulted in a single chip on a 6-inch silicon wafer.

Block	Width (μm)	Height (μm)
A	7766	1000
B	9216	1000
C	1000	1000
D	1351	9216
E	9216	9216
F	1000	9216
G	1351	2825
H	9216	2825
I	1000	2825

Figure 8.9. Size of each stitching block (in μm).

8.5 Architecture Design

Similar analog signal chain (ASC) architecture developed for prototype test chips was adapted for the 4Kx4K CMOS APS image sensor. Most of the ASC blocks such as amplifiers, decoders, and control logic were inherited from the prototype chip. Major difference in the 4Kx4K ASC architecture was that the mode of front-end sampling circuit was changed from voltage mode to charge mode. It was done by removing the front sampling capacitor and the PMOS source follower circuits of the prototype ASC architecture. This was done to reduce the power consumption of 4Kx4K imager. Also, the output sample and hold circuits were removed and sizes of the capacitances (C_C and C_F) were increased for better noise performance. A test structure was also added to test the readout channel gain from pixel photodiode node to the analog output pads. CMOS analog switches were used in the 4096x4114 image sensor architecture. Schematic of the modified analog signal chain (ASC) architecture is shown in Figure 8.10.

8.6 Functional Definition of Stitching-Blocks

After the architecture was designed, each stitching block sizes were determined and functional definitions of the blocks were identified. They are listed in Table 8.1. Most of the chip blocks contain power pads, routing resources, covered and open pixel arrays in various sizes.

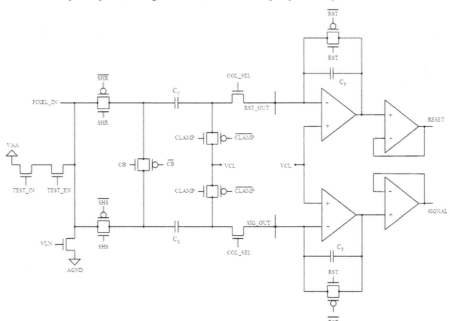

Figure 8.10. Analog signal chain (ASC) architecture of 4Kx4K CMOS image sensor.

Table 8.1. Functional definition of the stitching blocks.

Stitching-Block Name	Block Functional Definition
A	• Digital Power Routing • Covered/Open Pixels
B	• Pixel Power Supply Pads • Pixel Power Routing • Covered/Open Pixels
C	• Pixel Power Routing • Covered/Open Pixels
D	• Row Decoder/Drivers • Digital Input Pads • Digital Power Pads • Covered/Open Pixels • Signal/Digital Power Routing
E	• 512 x 512 Pixel Array
F	• Analog Power Pads • Power Routing • Covered/Open Pixels
G	• Analog/Digital Power Pads • Analog Signal Pads • Digital/Analog buffers • Row Clock Generator • Power/Signal Routing • Covered/Open Pixels
H	• Analog/Digital Input/Output Pads • Analog Signal Chain (ASC) • Column Decoder/Driver • Global Amplifiers • Column Timing and Control Logics • Analog/Digital Power Pads • Signal/Power Routing • Covered/Open Pixels
I	• Analog Input Pads • Analog/Digital Power Pads • Power routing • Covered/Open Pixels

Most of the digital processing and input/output operations were done on the left side of the chip. Right side was designated for analog operations, analog power inputs, and routings. Top size of the chip was used for pixel array's power inputs and routings. Most of the signal processing and signal routing were took place at the bottom blocks. Corner blocks were used for global signal routing and buffering purposes. All the side and corner blocks contain various size open and covered pixel arrays to make block transitions seamless as shown in Figure 8.11.

Parallelism was adapted in the readout of the pixel array. Pixel array was read through eight independent ASC channels that can be selected externally by enable signals.

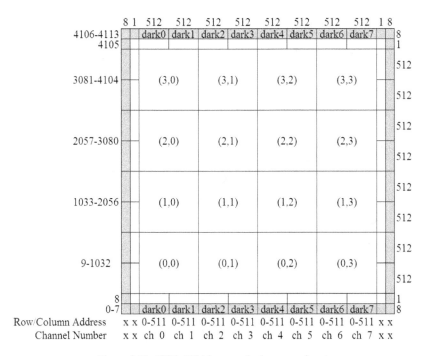

Figure 8.11. 4096x4114 imager pixel array and sector map.

Control and timing signals were generated locally in each block-H except for the global clock signals. This way catastrophic chip failure was avoided.

Seven columns of covered (dark) pixels were placed left and right sides of the pixel array. Eight rows of dark pixels were placed at the top and bottom sides of the pixel array, too. Open pixel transition between side stitching blocks and the pixel array block is achieved by placing an uncovered row and column of pixels before the stitch line borders on all sides. Based on these numbers, total number of pixels in the imager was 16,916,768 (4112x4114). First and last eight of the pixel array were not readable and used as electronic and optical barriers. They work like the other pixels but no readout channel was assigned to them. Thus total readable number of pixels was 16,850,944 (4096x4114). Because 16 of the rows were covered (dark), total number of imaging pixels was 16,785,408 (4096x4098).

Image sensor pixel array was divided artificially into sixteen 1024 by 1024 (1Kx1K) sub-sectors as shown in Figure 8.11. This was done to reduce the complexity of the testing system design and software that is running on PC. Testing software was designed to read one sector at a time while two channels are selected at any time.

199

8.7 Chip Timing Design

Timing diagram of 4096x4114 image sensor was designed according to the defined architecture and system integration issues. A region of interest could be read using external channel select signals (ch_i_en). This region of interest size could be as small as 521 x 1. Full frame and sector read timing diagrams are shown in Figure 8.12a, and Figure 8.12b, respectively.

Frame read operation starts with enabling the channels that are going to be read as shown in Figure 8.12a. For full frame, all the channels are enabled right after row pointer was initialized. Next, first row pixel signals are sampled onto column sample and hold circuits. First eight rows of pixels were dark pixels and used for signal processing purposes. After the row sampling completed, column address decoders of each channel were initialized. Since all the channels are enabled in full frame readout, sampled pixel signals of each channel were shifted out one by one during column read (CR) period. Total 16 analog outputs (8 for signal, 8 for reset) were active and available during column read period for off-chip ADC or PC frame grabber for digitizing the pixel signals. After all of the columns in the channel between 0 and 511 were shifted out, next row is sampled onto the column sample and hold circuits. After last row (row_4113) was read, next frame read operation starts. It is not shown on the timing diagram but, reset pointer is also running ahead of read pointer setting integration time for the pixels.

Sector read operation was shown for sector (2,2) in Figure 8.12b. Sector read operation timing is same as the full frame read until the end of the 7th row in where the dark rows end. Between the 8th row and the start of the sector address (2057th row), none of the rows are read (NR). They are only accessed, and the pixels are got reset for definite integration time period. Between sector start (2057th row) and the end address (3080th row) normal row read and column shift operations are performed. Between the sector end address (3080th row) and the last 8th dark row start address (4106th row), none of the rows are read again. After the last 8th dark row has been read, new frame read operation starts. Column and row read timing diagrams were same as the prototype chips reported in chapter 5. Because, most of the circuit blocks used in prototype chips were used in the 4Kx4K imager.

Row initialization, column initialization, column read, and row read time periods were set to 16, 16, 8, and 256 master clock cycles, respectively. Master clock frequency was chosen to be 24 MHz. Based on these numbers; frame rate of full-frame readout (4096 x 4114) was 1.24 FPS. If a sector read operation is performed, than the frame rate increases. Frame rates versus sector column sizes for various master clock frequencies are shown in the Figure 8.13 for the 4096x4114 device. For full-frame readout, an external 8 channel multiplexing, 24MSPS discrete ADC or a PC frame grabber is required.

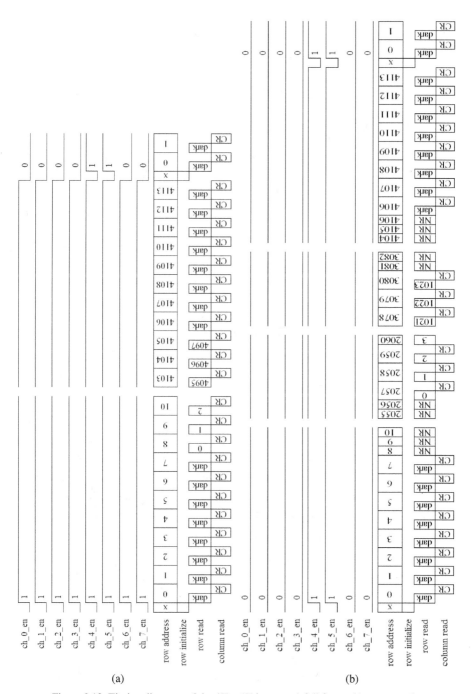

Figure 8.12. Timing diagram of the 4K x 4K imager; a) full frame, b) sector read.

201

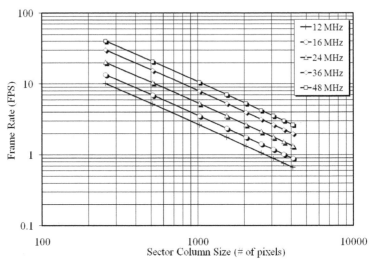

Figure 8.13. Sector column size versus frame rates for various master clock frequencies.

8.8 Detailed Stitching Block Design

After detailed stitching block definitions, functions and dimensions were identified, detailed design of each sub-block was done. Each of the blocks is described on designed block layouts in the following sub-sections.

8.8.1 Block A Design

Block A forms the upper left corner of the imager. It composes of one pixel array power pad (VAA_PIX), one analog ground (AGND) pad, nine shift register type row decoders and their driver circuits, and 8 x 9 array of pixels. The pixels were covered with top metal layer to form top-left side of the dark pixel rows and columns with the exception of the bottom right pixel. That pixel was not covered. The row decoders and drivers were used for addressing physical rows between 4105 and 4113. Power of the decoder and drivers were brought in from the stitched block below. Block A could be stitched to block B on the right side, and to the block D on the bottom side as shown in Figure 8.14.

8.8.2 Block B Design

Block B makes up of the middle top section of the imager. It was repeated 8 times on x-direction. Each Block B composes of one pixel array power pad (VAA_PIX), one analog ground (AGND) pad, and 512 x 9 arrays of pixels. Eight out of nine rows were covered with top metal layer to form top dark row of pixel array. The bottom row is an open or imaging row. Block B was

202

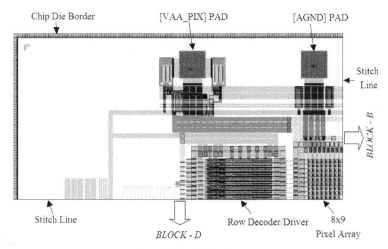

Figure 8.14. Block A layout.

designed to be stitched on three sides. It could be stitched to block A or block B on the left side , block B or block C on the right side, and block E on the bottom side of the layout as shown in Figure 8.15.

8.8.3 Block C Design

Block C forms the top right corner of the imager. It contains 8x9 array of pixels, one ground (AGND) pad, and power routings. Again, except for bottom-left of the pixel, all other pixels in 8x9 arrays are covered with top metal layer forming dark pixels. Block C could only be stitched to block B on the left, and block F on the bottom side of the block layout as shown in the Figure 8.16.

8.8.4 Block D Design

Block D forms the middle left side of the imager. It contains 512 row decoders, 512 row drivers, 8 x 512 array of pixels, digital signal and power pads, and routing resources. Block D could be stitched to block A or block D on top, to block D or block G on the bottom side, and to block E on the right side as shown in the Figure 8.17. Fifteen digital input pads were placed in block D.

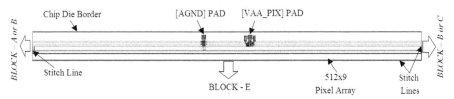

Figure 8.15. Block B layout.

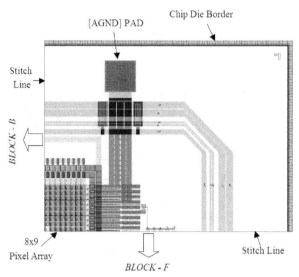

Figure 8.16. Block C layout.

These digital input signals were routed only towards the bottom of the block D as seen in Figure 8.18. If block G placed at the bottom of block D, then these input signals were passed to block G. If the block D placed at the bottom, than they were connected a load capacitor and left unconnected.

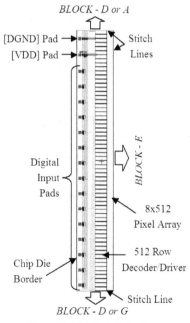

Figure 8.17. Block D layout.

204

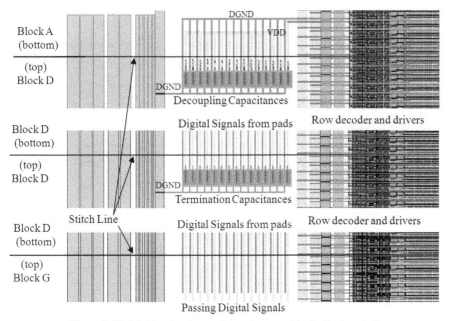

Figure 8.18. Stitching border transitions between blocks D, A, and G.

If block A is on top side, capacitances on the top side of the block D are connected to supply rail in block A, turning capacitances in block D into power supply decoupling capacitors. This way, only the block D placed above block G brings in digital input signals to the chip. This approach reduces the number of pads that are required to be placed in the block G. There were also one digital ground (DGND), and one digital power (VDD) pads placed in each block D.

Static type shift registers were used as row decoders. This type of decoding scheme is very suitable for very long stitched decoders with less number of signals running on the stitching borders. Row drivers were based on a digital tapered buffer.

8.8.5 Block E Design

Block E composes of 512 x 512 pixel array. This block is the one repeated 64 times during fabrication to form main imaging array of 4096 x 4096. A four side stitchable pixel layout is used in this block. Pixel layout is shown in Figure 8.7. Block E could be stitched to block E or block B on top, block E or block F on right, block E or block D on left, and block E or block H on the bottom side of the block. Stitch line between surrounding blocks are seamless due to the fact that every block surrounding the block E contains open, imaging pixel on its block E side.

205

8.8.6 Block F Design

Block F forms the middle right side of the imager. It contains 8x512 array of pixels on the left, two power pads (VAA and AGND), 5 analog input pads, and signal routing. First column on the left (block E side) contains open imaging pixels while the rest of the columns were covered with top metal layer to form the dark pixels. Analog pad outputs were routed towards the bottom of the block same as the block D. They were connected to grounded capacitances which are placed at the top of the block F layout if block F was placed below. Other vice, the analog input signals were passed to block I. Block F could be stitched to block F or block C on top, to block F or block I on the bottom, and only to block E on the left side of the block as shown in Figure 8.19.

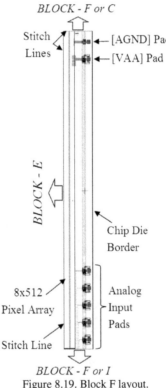

Figure 8.19. Block F layout.

8.8.7 Block G Design

Block G forms the lower left corner of the imager. It contains row timing generator, 8x9 array of pixels, nine row shift register and row drivers, tapered digital buffers, analog, digital and power pads, and signal routings as shown in the Figure 8.20. Block G can be stitched to block D on top, and to block H on right side.

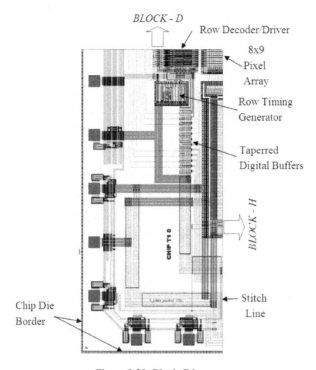

BLOCK - D

Row Decoder/Driver

8x9
Pixel
Array

Row Timing
Generator

Taperred
Digital Buffers

BLOCK - H

Chip Die
Border

Stitch
Line

Figure 8.20. Block G layout.

8.8.8 Block H Design

Readout channel is placed in block H. Correlated double sampling (CDS) was performed on-chip in ASC block and the difference of the pixel signals were sent off chip in analog format to be digitized. Sampling of a selected row was done simultaneously on all readout channels. Only selected readout channels shifted out their column sample and hold (CSH) contents. Each readout channel or block H composes of 512 of analog signal processor (ASP), 512 column decoders, pseudo-differential analog buffers, timing and control signal generator for column decoder and ASC, digital buffers, input-output pads, and 512x8 array of pixels as shown in Figure 8.21.

To allow smooth transition between channel to channel, column pitch was reduced to 17.8μm as oppose to the pixel pitch of 17.8μm. This way, a 102.4μm open space was left between adjacent block H layouts. In this space, globally distributed signals such as SHS, SHR, and CLAMP signals were routed locally in each block H. This way, a defect that would short any of these global signals in column layout would make only the channel or block H in-operational, not the whole chip. This improves the yield which is the major concern of very large format image sensor IC design. Layout of the block H or one readout channel is shown in Figure 8.22.

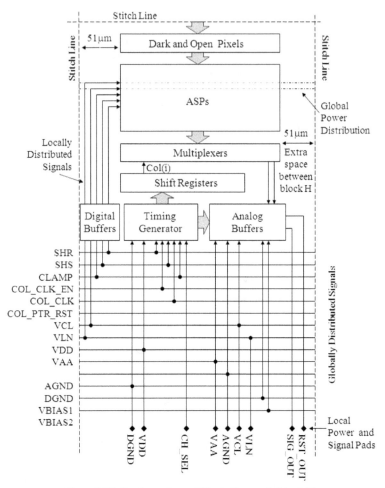

Figure 8.21. Readout channel block diagram in block H.

Some of the sensitive signals such as biases, and analog references were brought in readout block through local pads, and routed from left of the chip to the right. This way, uniformity of the supply and bias voltages were achieved. This resulted in less imaging artifacts and better power distribution. Less sensitive and slow digital signals were buffered at the bottom left side of the chip block (Block G), and routed left to right in block H without having local pads. To reduce the loading on this long distribution wires, each distributed signal was buffered locally in block H. One timing generator placed in each readout block. It composes of column clock generator and controller. This block is inherited from the prototype test chip. A digital control input (CH_SEL) for each channel was used for enabling timing generator during readout phase so that user could

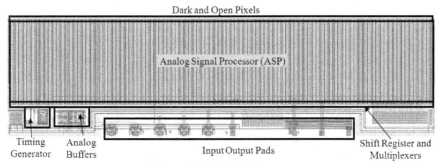

Figure 8.22. Block H layout.

be able to choose which block to read and which block to shut down for power saving. Signal gain between photodiode node and analog output pads was set to unity by arranging the capacitance values of the analog signal chain. Signal chain works linearly in 2-volt effective pixel signal range.

8.8.9 Block I Design

Block I forms the bottom right side of the 4096x4114 imager. It contains 8x9 array of pixels, power and analog input pads, and signal routings as shown in Figure 8.23. It can only be stitched to block F on top, and to block H on the left side.

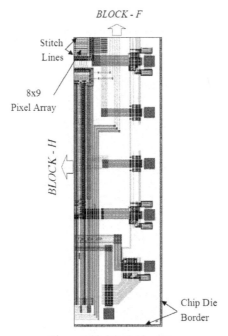

Figure 8.23. Block I layout.

8.9 Chip Assembly and Packaging Design

Since there were no standard IC package available that could hold 76mm x 77mm die, a custom packaging solution was developed for the 4096x4114 sensor. Block diagram of the packaging PCB is shown in Figure 8.24. Packaging PCB has two headers bringing the control and other necessary signals for the 4096x4114 device and other IC's. Since the analog output drivers of the 4096x4114 sensor are not strong enough to drive heavy loads, extra analog switch/buffer IC's were used on the packaging board. A 4-to-1 programmable analog buffer IC was chosen for this purpose. Two out of 8 analog differential output of the 4096x4114 device was selected for buffering. Gain of the analog buffer IC's was set by jumpers. Gain can be selected 1, 2, 4, and 8 times. All the power, ground and bias voltages were generated outside the packaging PCB.

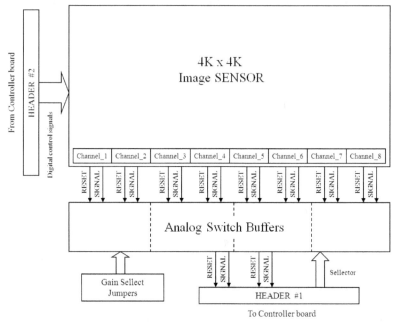

Figure 8.24. Packaging PCB (PPCB) block diagram.

Printed circuit board (PCB) layout of packaging solution is shown in Figure 8.25. Packaging PCB compose of a large sit plane and 156 bonding pad fingers for 4096x4114 sensor. These fingers and the sit plane were plated with gold for better packaging performance and sticking the wire bonds. 4096x4114 die was glued on the sit plane by a conductive epoxy and connected to analog pixel ground plane in the PCB. Extra holes were drilled on the PCB to hold the front glass to protect the die after packaging. The 4096x4114 sensor was mounted on the packaging PCB. Pictures of front and back side of the packaging PCB after assembly is show in Figure 8.26.

Controller board designed such a way that packaging PCB could be plugged onto the controller board, as shown in Figure 8.27. A projector was used to image test patterns on the imager.

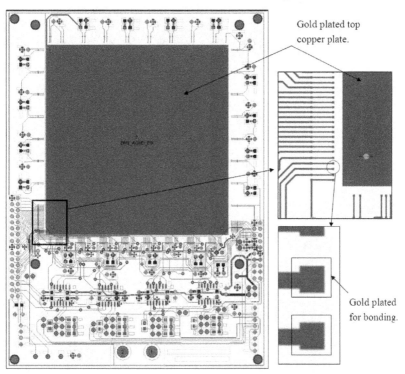

Gold plated top copper plate.

Gold plated for bonding.

Figure 8.25. PCB for packaging 4096x4114 image sensor.

Figure 8.26. Front and back side of the Packaging PCB after assembly.

Figure 8.27. Controller PCB and packaging PCB connection.

8.10 Demonstration and Test System Design

A demonstration and characterization system and viewer software was developed for the 4096x4114 imager.

8.10.1 Controller Printed Circuit Board (PCB) Design

New printed circuit boards (PCB) were designed for the sensor. One PCB was developed to package the 4096x4114 device (packaging PCB), and the other board was to generate controller signals for the packaging PCB and to interface to a computer (controller PCB). Block diagram of the controller PCB is shown in Figure 8.28.

212

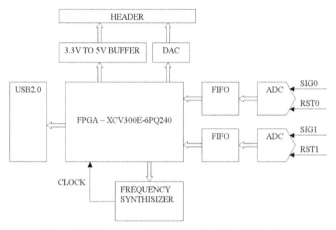

Figure 8.28. Demonstration system for 4K x 4K device.

The key part of the design is the Field Programmable Gate Array (FPGA), which controls all other components on board. Based on the clock from the clock synthesizer, FPGA generates digital signals for the sensor. With the clock synthesizer, the clock rate could be adjusted from 20.83MHz to 500MHz. This makes it possible to change the frame rate and integration time easily. Since the sensor is working at 5 volt, while the FPGA is working at 3.3 volt, some level shifter buffers were inserted.

The whole image size is 4096x4114, where the upper 8 rows and the lower 8 rows are black rows. These black rows were used to do black level calibration to get rid of the column wise fix pattern noise. The image is divided to 4x4 blocks. User can choose one of them to be outputted from the sensor in a frame. Once the sensor gets the correct timing, it will output two differential signals, each of them is corresponding to the adjacent 512x1040 block, so that we can get 1024x1040 pixels.

The 14-bit analog to digital converter (ADC) was running at the pixel clock (1/16 of the master clock). The image data was going directly to two first-in-first-out (FIFO) buffers, along with a frame valid and a line valid signal. So, when the FPGA retrieves data from the FIFO, it knows the start and the end of a frame and a row.

The universal serial bus (USB2.0) controller is responsible for the communication with a computer. It can achieve up to 480 megabits per-second data transfer rate. Data from the FPGA was fed to the USB, and then sent to a computer. Meanwhile, control registers in the FPGA could be reached through the USB port. With these registers, the frequency synthesizer, on board digital to analog converters (DAC), and other parameters were controlled.

8.10.2 Controller PCB Design Considerations

Frame Rate: The goal was to get more than 1 frame (1024x1024 pixels) per second. When the master clock was 24MHz, the pixel clock was set to 24/16=1.5Mhz in one operation setting. One frame consists of row init, row reset, signal sample and hold, row read, and column read timing sequences. At 12 MHz the frame rate was about 1.25 FPS.

Integration Time: The integration time (the time that pixels were exposed) should be changeable to get decent images in bright and dark light. An easy way is to change the master clock. The clock synthesizer was design to provide programmable master clock rates between 20.83 and 500MHz.

Analog-to-Digital-Converter (ADC): A 14-bit ADC was chosen for converting differential analog signals coming from the imager. ADC was running up to 65 mega samples per second, with the input range of differential ±1 volt. According to its spec, the differential non-linearity (DNL) was ±1 least significant bit (LSB).

Frame Buffer Size: The FIFO (buffer) size of each channel was 512Kx16bits. So a whole 1024x1024x14bit image could be stored on test board.

Programmable Controller (FPGA): A Xilinx FPGA was used in the design. It has more than 400,000 system gates and 158 I/O pins. Very high speed integrated circuit hardware description language (VHDL) programming language was used to generate timing signals in FPGA.

Digital-to-Analog Converter (DAC): 10 and 12-bit series DACs were used for generating bias voltages for the imager. DACs provided rail-to-rail programmability of the bias voltages with better that 0.5 LSB linearity.

8.10.3 FPGA Controller Blocks

FPGA control blocks are shown in Figure 8.29. Dac_ctrl is the interface controller block for the DAC. It had a three-wire serial interface. This module receives commands from Reg_ctrl, and generates proper timing for the DAC. Read_fifo and Write_fifo blocks manage the FIFO chips on the test board for read and write operations. Freq_syn block controls frequency synthesizer chip. It also provides low voltage positively referenced emitter coupled logic (LVPECL) signal interface. Reg_ctrl is the register control module. When the I^{2}C engine receives data, it will pass the data to this module to store the values. Or when the I^{2}C engine wants to read registers, it will get data from this module. Moreover, this module controls other modules, according to its registers. Row_col manages sensor timing. Clock_gen manages clock distribution of each block deriving clock from the 48MHz clock input. Write_usb generated proper signals for USB controller chip.

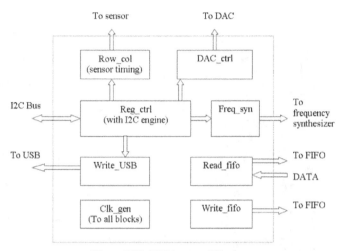

Figure 8.29. FPGA control blocks.

8.10.4 User Interface Program Design

A screen shot of the test board control and image viewer software is shown in Figure 8.30. This software controls test board parameters through PC's USB2 port. These parameters are the bias voltages for the 4096x4114 imager (Vbias1, Vbias2, Vcl, Vln, Vtest_in), channel select signals, and FPGA timing generator's master clock frequency. As a image viewer software, it allows user to choose which sector of the 4Kx4K imager to read between 0,0 and 3,3. It also allow user to adjust offset and gain of raw image data using the following formula.

$$\text{Image data} = \text{Gain} * (\text{Raw data} - \text{Offset}) \qquad [8.1]$$

Offset and gain set to 8000 (LSB), and 2 respectively. It allows user to save image in bitmap (.BMP) or raw binary format (.RAW). It informs user how many frames were read, display and camera frame rates at the top section of the window. When the program started, it communicates the default setting like biases, sector, and clock frequency information to the board. Picture in Figure 8.30 was taken while the 4096x4114 device was testing. A projected test image on the sector (1,1) at 12MHz master clock frequency and 1.2 FPS frame rate was imaged.

8.11   Measurement Results

Characterization was done in room temperature. A projector was used to drop uniform light on a section of the imager. Since the projector exit opening was very small comparing the size of the device, certain sub-array of the 4096x4114 imaging array was used for characterization. For characterization a seperate in-house program was used to collect and process images.

215

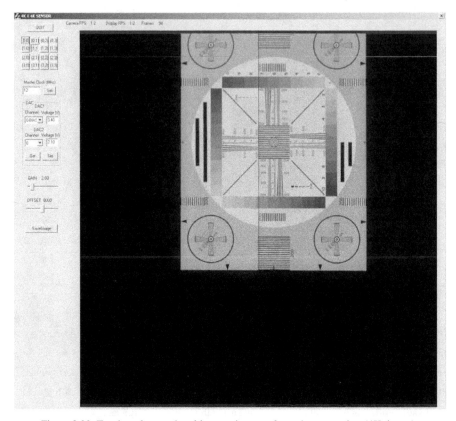

Figure 8.30. Test board control and image viewer software's screen shot (4Kviewer).

8.11.1 Integration Time Control

Integration time of the imager was adjusted by changing the master clock of the controller board through the 4Kviewer program. Integration time versus clock frequency is plotted in Figure 8.31. Default master clock frequency was 24MHz leading to 0.395 second integration time. 4096x4114 device was read 2 channel at a time leading to 1Kx1K block images. These images was buffered on test board and transferred to the PC through USB 2.0 bus interface. 4Kviewer software then shows these data on the PC monitor in 1Kx1K formats.

An image can be dropped on the 4096x4114 device by large lens, and the array can be scanned by 4Kviewer software. Each 1Kx1K sections of the array can be saved in bitmap (.BMP) file format to reproduce the whole 4096x4114 image. Test images taken through projector and through a large enough lens were presented at the last section of this chapter.

216

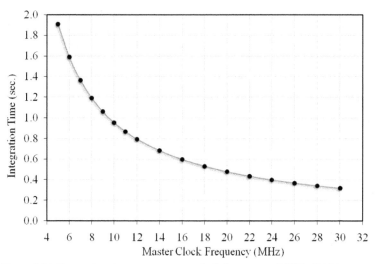

Figure 8.31. Integration time versus master clock frequency of 4096x4114 imager.

8.11.2 Power Consumption

Power consumption of the 4096x4114 device is mainly depending on the column source follower current consumption that is set by the VLN bias voltage. Power consumption of the chip alone at 12MHz clock frequency is shown in Figure 8.32. Typical operation voltage of VLN is between 1.0 and 1.2 volt leading to 500 to 750 mWatts of power consumption. All measurements were done at 1.05 volt VLN voltage, and 550mWatts power consumption.

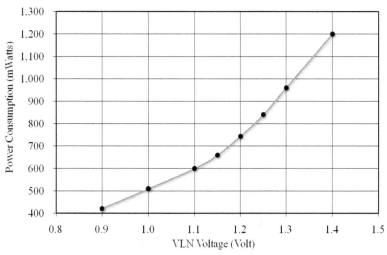

Figure 8.32. Power consumption versus VLN bias voltage of 4096x4114 imager.

8.11.3 Dark Current

Dark current was measured by adjusting the integration time of particular sector. Integration time swept from 0.35 to 0.6 second and the frame averages were calculated. Measured dark current voltage is shown in Figure 8.33. It was 57.4 mVolt/sec, or 25.3Ke-/second. Expected dark electron rate was 24.0Ke-/sec.

The increase on the dark current was related to the heat management of the packaging PCB and the controller board. Controller board was consuming 0.98 amper on 7.0 Volt power supply leading to about 7 Watts of power consumption. So the controller board was heating up the immediate surrounding where the 4096x4114 imager chip located. This temperature increase on the packaging PCB of the 4096x4114 device die resulted in the dark current increase.

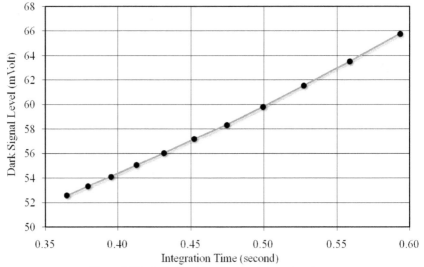

Figure 8.33. Dark signal level versus integration time.

8.11.4 Sensitivity and Linearity

Sensitivity of the imager was measured by stepping the light source from complete darkness to full well illumination in precisely measured increments. At each illumination level the mean values of the pixels were calculated and recorded. A sharp pass band green filter (550+40nm) was used at the output of the light source. Light level was measured in foot/candle, and converted to lux. Measurements were performed at 24Mhz clock frequency leading to 0.395 second integration time. For each pixel, mean pixel value in volt versus exposure in lux*sec were plotted and regional sensitivity of the pixel was calculated by determining the slope of the plots.

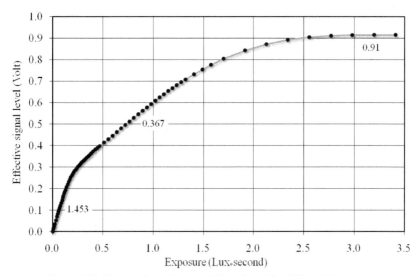

Figure 8.34. Measured sensitivity and linearity of the 4096x4114 imager.

Two region of operation were observed on the sensitivity curve as shown in Figure 8.35. First section was the high conversion gain operation leading to low pixel capacitance and high sensitivity of 1.453 Volt/lux.sec. The second section was low sensitivity of 0.367 Volt/lux.sec because of a low pixel conversion gain. The pixel saturation level was 0.91 volt. Imager had a 0.7 volt linear range.

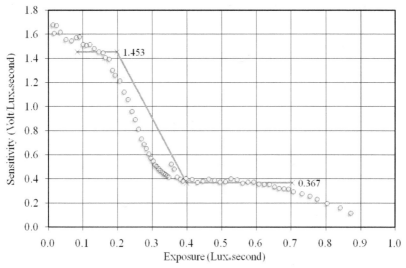

Figure 8.35. Sensitivity measurement results of the 4096x4114 imager.

219

8.11.5 Analog Signal Chain (ASC) Gain

Analog signal chain (ASC) gain from pixel floating diffusion node to external ADC input was measured by using the test structures placed in each column. The gain of the signal chain from pixel to chip output was designed to be unity. The external amplifier IC's gain was set to two (2). Thus we expected to have an overall gain of 2 from pixel floating diffusion to external ADC inputs. the measurement result at nominal bias and clock conditions is shown in Figure 8.36. Measurement showed that at nominal biasing levels, the signal chain had 1.707 Volt/Volt gain and 18mVolt of positive offset. This gain was confirmed at different biasing conditions at 24 MHz master clock frequency.

8.11.6 Conversion Gain and Full-Well Capacity

Conversion gain was measured at nominal bias settings. Photon transfer curve of the imager is shown in Figure 8.37. Pixel capacitance transition from low capacity to high capacity could be observed on the photon transfer curve. Up to 0.2 volt effective pixel voltages, the pixel NMOSCAP capacitance was turned off fully leading to low pixel capacitance. Between 0.2 volt and 0.4 volt, NMOSCAP capacitance turns on and becomes fully ON at 0.4 Volt leading to a large total pixel capacitance. Noise floor of the 4096x4114 image sensor chip was measured. It was 240 electrons. Measured pixel conversion gain was 2.27μVolt/e- for signal level below 0.2volt, and 0.23μVolt/e- for signal level larger than 0.4 volt. In between 0.2 and 0.4 volts, pixel conversion

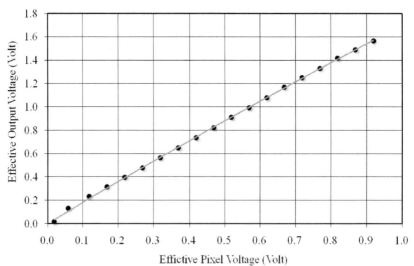

Figure 8.36. Analog signal chain (ASC) gain from pixel photodiode node to ADC input at nominal biasing conditions for 4096x4114 imager.

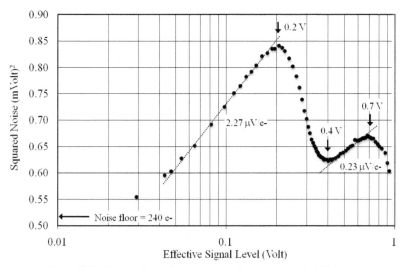

Figure 8.37. Conversion gain measurement result at nominal biasing.

gain changes between two conversion gain values. Equivalent pixel capacitances at different effective signal levels were calculated based on the linear signal range and the conversion gain. It changes between 70fF (C1) and 700fF (C2).

Pixel full-well capacity versus the signal level is shown in Figure 8.38. Measured pixel full-well capacity was 1.35 million electrons. Expected full-well capacity was 1.05 million electrons Noise floor of 240 electrons with 1.35 million electron pixel full-well capacity resulted in 75 dB interscene dynamic range for the 4096x4114 imager.

8.11.7 Pictures Taken with the 4Kx4K Image Sensor

It was determined that in all the packaged parts channel 0, 1, 6, and 7 do not respond to light. Other channel outputs works with individual offsets as seen clearly on the reproduced image in Figure 8.39. These offsets can be fixed with signal processing methods after the images were captured by a particular device. Dead rows and columns, and point defects were easily seen on each sector due to the fabrication defects as shown in Figure 8.40. On the stitching line borders, one or two rows and column are not functioning properly. These might be caused by a single pixel defect on the stitch line because of stitching imperfection. These image defects like channels offsets, dead rows, and columns could be fixed by means of signal processing in PC before showing the image on the monitor. Focused sections of the captured images shown in Figure 8.39 and Figure 8.42 are shown in Figure 8.40, Figure 8.41, and Figure 8.43, respectively.

221

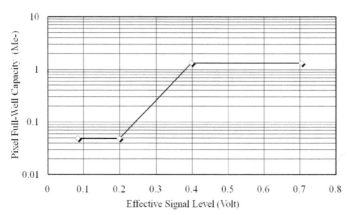
Figure 8.38. Effective signal level versus pixel capacitance.

ch0   ch1   ch2   ch3   ch4   ch5   ch6   ch7

Figure 8.39. Full-frame reproduced image of imaging array and a test pattern by an image projector on the stitching borders. (4096x4114)

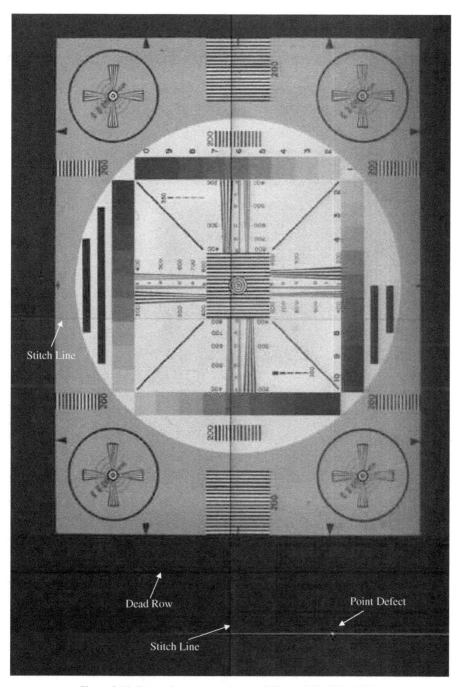

Figure 8.40. Focused test pattern image of Figure 8.39 (976x1428).

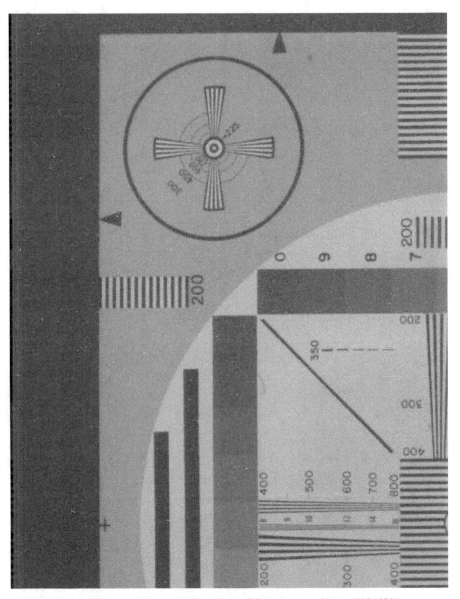

Figure 8.41. Focused upper-left quadrant of the test pattern image (512x654).

Figure 8.42. Reproduced raw scene image under poor lighting and focus (4096x4114).

Figure 8.43. Focused sections of the image shown in Figure 8.42. (sized raw images).

8.12 Case Study Summary

CMOS image sensor design and technology evaluation methodologies for fabless CMOS image sensor design companies for first silicon success were reviewed. A stitching design methodology was proposed to design very large die size CMOS image sensor. It was successfully implemented in a case study. In the case study a CMOS APS image sensor with more than 16 million pixels and 1 million electron pixel full-well capacity was designed in a 0.5μm, 5volt, 2P3M, CMOS CIS process with stitching option. A new hybrid photodiode-photogate (HPDPG) CMOS APS pixel with 18μm pixel pitch was designed and used in the imager. By using stitching option, designed image sensor resulted in large single die CMOS image sensor occupying 76.05mm by 77.55mm (5898 mm^2) on 6-inch silicon wafer. Imaging array area was 73.73mm x 74.05mm (5460 mm^2). New packaging and system integration methods were also proposed and used during design to demonstrate feasibility and functionality of a wafer scale, single die, and 16.85 million pixels (4096x4114) CMOS image sensor with more than 1.35 million electrons pixel full-well capacity. General specification and the measurement results of the imager are listed in Table 8.2. Quantum efficiency of the imager was not measured but extrapolated from the test pixel data. 4096x4114 imager resulted in 75 decibel (dB) dynamic range even though the noise floor of the imager in the test system was 240 electrons.

Table 8.2. Specifications and measurement results of 4096 x 4114 image sensor.

Process	0.5μm, 2P3M CMOS with stitching option	
Pixel Size	18 x 18	μm^2
Pixel Type	Hybrid Photodiode-Photogate (HPDPG) APS	
Total Pixel	16,916,768 (4112x4114)	
Effective Pixels	16,850,944 (4096x4114)	
Pixel Array Dimensions	73.73 x 74.05	mm^2
Chip Die Dimensions	76.05 x 77.55	mm^2
Frame Rate	1.25	FPS
Conversion Gain	2.27	μV/e-
	0.23	μV/e-
Pixel Full-Well Capacity	1,350	Ke-
Sensitivity	1.45	Volt/Lux-sec.
	0.37	Volt/Lux-sec.
Quantum Efficieny	24*	% @ 390nm
	46*	% @ 550nm
Dark Current	57.4	mVolt/sec.
	25.3	e-/sec.
SNR (max)	61.3	dB
Read Noise	240	e-
Dynamic Range	75.0	dB
Supply Voltage	5.0	Volt
Power Consumption	550	mWatts

* : Extrapolated

227

8.13   Conclusion and Future Prospect

Emergence of CMOS active pixel sensor (APS) technology as a candidate to replace CCD technology in almost all image sensor applications has opened new research grounds for high performance image sensor design. Design and performance issues for large format CMOS APS image sensors were addressed in this monogram.

8.13.1 Conclusions

Investigations were focused on four main problems of CMOS APS image sensor design. They were; CMOS APS pixel development with very large pixel full-well capacity for high dynamic range imaging, quantum efficiency improvement in blue spectrum (400-450nm), large format image sensor development that could exceed standard CMOS reticle size of 20mm by 20mm, and development of CMOS image sensor design and evaluation models and methodologies for first time silicon success.

Pixel full-well capacity improvement was achieved by two proposed methods. In first method, CMOS APS pixel photodiode peripheral was increased by opening few circular openings on the pixel photodiode diffusion. This method was called pixel peripheral utilization method (PPUM). It results in photodiode parasitic capacitance increase only if the unit peripheral capacitance of diffusion region is comparable with the unit area capacitance. It was found from evaluating few commercial CMOS process parameters that, in general, this condition is satisfied by CMOS process technologies that have larger than 0.5 μm minimum feature sizes. Thus, pixel peripheral utilization method was demonstrated on a 0.5 μm, 2P3M, 5V, CMOS process by designing test pixels in a prototype CMOS APS image sensor. At least 25% full-well capacity improvement was demonstrated in an 18 μm pixel pitch by adding seventeen 1.6μm diameters circular openings on the photodiode region. In second pixel full-well capacity improvement method, an in-pixel NMOSCAP was used in a new CMOS hybrid photodiode-photogate (HPDPG) APS pixel. New APS pixel was achieved more than 4 million electron (Me-) pixel full-well capacity in an 18μm pixel.

Pixel quantum efficiency improvement, especially in blue spectrum (400-450nm), was achieved by improving lateral collection efficiency of photodiode type CMOS APS pixel. Lateral collection efficiency was improved by increasing photodiode peripherals. At least 10% quantum efficiency improvement was demonstrated at blue spectrum by placing circular openings on photodiode of APS pixel. This was the same method that was used for pixel full-well capacity improvement. Thus, proposed pixel peripheral utilization method improves both quantum efficiency at blue spectrum and pixel full-well capacity in general.

228

A number of CMOS pixel property models were developed for CMOS image sensors. Tree CMOS pixel technologies were investigated for modeling; CMOS passive pixel technology, CMOS photodiode type active pixel sensor (APS) technology, and CMOS photogate type active pixel sensor (APS) technology. A hybrid threshold voltage equation was derived for minimum feature size sub-micron CMOS devices. It was used in electrical parameter modeling of the CMOS pixels during technology evaluation and image sensor design processes. Modeled APS pixel electrical parameters were; photodiode reset level, pixel signal range, photodiode and photogate type pixel full-well capacity, and passive pixel column parasitics. A pixel design and physical modeling tool called pixel physical property table (PPCT) was proposed. By using PPCTs, physical properties of CMOS pixel technologies were investigated and pixel layouts were optimized. CMOS pixel pitch and pixel fill factor were modeled.

Developed physical and electrical models were used in proposed technology evaluation methodology for fabless CMOS image sensor design companies. A new stitching design methodology was developed for very large die size image sensor development. It was demonstrated in a case study. A 16.85 Million pixel (4096 x 4114), wafer scale, CMOS APS image sensor with 1.35Me- full-well capacity was successfully designed, fabricated, and demonstrated using proposed stitching design methodology in a 0.5 μm, 5 volt, 2P3M, CMOS process with stitching option. Designed CMOS image sensor was fabricated on a single 6-inch silicon wafer. Unit area full-well capacity is defined by the ration of total pixel full-well capacity at saturation to pixel area in micrometer. It is 4167 e-/μm^2 for the imager. Microphotographs of designed 4096x4114 pixel CMOS APS image sensor on 6-inch silicon wafer before and after dicing are shown in Figure 8.44 and Figure 8.45, respectively.

8.13.2 Future Prospect

Number of methods that has been used in CCD technologies for quantum efficiency improvement in UV/blue spectrum was not implemented. Those were the phosphor coating for UV enhancement and backside illumination of the image sensor after backside thinning. Both of these techniques require a major development and design effort on the process side. Thus, their implementation was left as future development areas for CMOS quantum efficiency improvement.

Another improvement was the noise floor of the image sensor. Because the designed sensors were using pseudo-differential readout channels, noise floor was not optimized accurately leading to one-order higher noise floor in the signal chains. Fully-differential signal chains are widely used in modern CMOS APS image sensor IC today. Their design and noise optimization techniques were left out in this work.

Figure 8.44. Photograph of 4096x4114 CMOS APS sensor on 6-inch silicon wafer before dicing.

Figure 8.45. Microphotograph of 4096x4114 CMOS APS sensor die after dicing.

REFERENCES

[Ackland96]   B. Ackland, A. Dickinson, "Camera on a Chip," *Digest of Technical Papers, IEEE International Solid-State Circuits Conference (ISSCC),* pp.22-25, 8-10 Feb. 1996.

[Amelio70]   G. F. Amelio, M. F. Tompsett, G. E. Smith, "Experimental verification of the charge couple device concept," *Bell Systems Technical Journal,* vol. 49, pp. 593-600, April 1970.

[Anemogiannis92] E. Anemogiannis, E. N. Glytsis, "Multilayer waveguides: Efficient numerical analysis of general structures," *IEEE Journal of Lightwave Technology,* vol. 10, Issue 10, pp.1344-1351, Oct. 1992.

[AtmelURL1]   "AT71201M Datasheet" http://www.atmel.com/

[Ay02]   S. U. AY, M. Lesser, E. R. Fossum, "CMOS Active Pixel Sensor (APS) Image sensor for Scientific Applications", *Proc. of SPIE,* vol. 4836, pp. 271-278, 2002.

[Ay05]   S. U. Ay, "Electrical Property Modeling of Photodiode type CMOS Active Pixel Sensor (APS) Pixels", 48th IEEE Int'l Midwest Symposium on Circuits & Systems (MWSCAS), Aug. 7-10, 2005, Cincinnati, Ohio, USA.

[Aw96]   C. H. Aw, B. A. Wooley, "A 128x128-pixel standard-CMOS image sensor with electronic shutter", *Digest of Technical Papers, IEEE International Solid-State Circuits Conference (ISSCC),* vol. 39, pp. 180-181, 1996.

[Balanis89]   C. A. Balanis, *Advanced Engineering Electromagnetics,* John Wiley & Sons, 1989.

[Beecken96]   B. P. Beecken, E. R. Fossum, "Determination of the conversion gain and the accuracy of its measurement for detector elements and arrays," *Applied Optics-OT,* vol. 35, no. 19, pp. 3471, July 1996.

[Berezin01]   V. Berezin "Small CMOS pixel design with single row line, " *IEEE Workshop on CCDs and Advanced Image Sensors,* pp. 64-67, Crystal Bay, Nevada, June 2001.

[Blanksby97]   A. J. Blanksby, M. J. Loinaz, D. A. Inglis, B. D. Ackland, "Noise performance of a color CMOS photogate image sensor," *IEDM Technical Digest, International Electron Devices Meeting,* pp. 205-208, 7-10 Dec. 1997

[Blanksby00]   A. J. Blanksby, M. J. Loinaz, "Performance analysis of a color CMOS photogate image sensor," *IEEE Transactions on Electron Devices,* vol. 47, no. 1, pp. 55-64, Jan. 2000.

[Blouke80]   M. Blouke, M. Cowens, J. Hall, J. Westphal, A. Cristensen, "Ultraviolet downconverting phosphor for use with silicon CCD image sensors," *Applied Optics,* vol. 19, no. 19, 1980.

[Boyle70]   W. S. Boyle, G. E. Smith, "Charge Coupled Semiconductor Devices," *Bell Systems Technical Journal,* vol. 49, pp. 587-593, April 1970.

[Bredthauer07]	R. Bredthauer, K. Boggs, G. Bredthauer, "A monolithic 111-M Pixel High Speed, High Resolution CCD," *2007 International Image Sensor Workshop*, pp. 170-173, Ogunquit, Maine, June 7-10 2007.
[Burkey84]	Burkey *et al.*, "The pinned photodiode for an interline-transfer CCD image sensor," *IEDM Technical Digest, International Electron Devices Meeting*, pp. 28-31, 1984.
[Chamberlain83]	S. G. Chamberlain, J. Lee, "A novel wide dynamic range silicon photodetector and linear imaging array," *Proc. IEEE Custom Integrated Circuits Conf.*, pp. 441-445, 1983.
[ChangJ94]	J. Chang, A. A. Abidi, C. R. Viswanathan, "Flicker noise in CMOS transistors from subthreshold to strong inversion at various temperatures," *IEEE Transactions on Electron Devices*, vol. 41, pp. 1965-1971, Nov. 1994.
[Cheng97]	Y. Cheng, M. C. Jeng, Z. Liu, J. Huang, M. Chan, K. Chen, P. K. Ko, C. Hu, "A Physical and Scalable I–V Model in BSIM3v3 for Analog/Digital Circuit Simulation," *IEEE Transactions on Electron Devices*, vol. ED-44, pp. 277-287, 1997.
[Cheng98]	Y. Cheng, M. Chan, K. Hui, M. C. Jeng, Z. Liu, J. Huang, K. Chen, J. Chen, R. Tu, P. K. Ko, C. Hu, *BSIM3v3 Version 2.0 User's Manual*, June 1998.
[Cheng03]	H. Y. Cheng, Y. C. King, "A CMOS image sensor with dark-current cancellation and dynamic sensitivity operations," *IEEE Transactions on Electron Devices*, vol. 50, no. 1, pp. 91-95, Jan. 2003.
[Ching02]	C. H. Ching, W. Shou-Gwo, Y. Dun-Nian, T. Chien-Hsien, L. Jeng-Shyan, C. S. Wang, C. Chin-Kung, H. Yu-Kung, "Active pixel image sensor scale down in 0.18 um CMOS technology," *IEDM Technical Digest, International Electron Devices Meeting, pp. 813-816, 8-11 Dec. 2002*.
[Cho00]	K.-B. Cho, A. Krymsky, E. R. Fossum, "A 1.2 V Micropower CMOS Active Pixel Image Sensor for Portable Applications," *Digest of Technical Papers, IEEE International Solid-State Circuits Conference (ISSCC)*, pp.114-115, 7-9 Feb. 2000.
[Cho01]	K.-B. Cho, A. Krymsky, E. R. Fossum, "A micropower self-clocked camera-on-a-chip," *IEEE Workshop on CCDs and Advanced Image Sensors*, pp. 12-15, Crystal Bay, Nevada, June 2001.
[Cohen80]	C. Cohen, "Color TV camera using CCD image sensor chips gets first sale," *Electronics*, pp. 78-80, Feb. 14, 1980.
[Cowens80]	M. Cowens, M. Blouke, T. Fairchild, J. Westphal, "Coronene and liumogen as VUV sensitive coatings for Si CCD image sensors: a comparison," *Applied Optics*, vol. 19, no. 22, pp.3727, 1980.
[DalsaURL1]	"FTF4052C Datasheet" http://www.dalsa.com/pi/
[Dickinson95]	Dickinson, *et al.*, *IEEE Workshop on CCDs and Advanced Image Sensors*, Dana Point, CA, 20-22 April 1995.
[Dickinson96]	Dickinson, *et al.*, *IEEE Workshop on CCDs and Advanced Image Sensors*, 1996.

[Dierickx96]   B. Dierickx, D. Scheffer, G. Meynants, W. Ogiers, J. Vlummens, "Random addressable active pixel image sensors," *Proc. of SPIE,* vol. 2950, pp. 1, 1996.

[Dyck68]   R. Dyck, G. Weckler, "Integrated arrays of silicon photodetectors for image sensing," *IEEE Transactions on Electron Devices,* vol. ED-15(4), pp. 196-201, 1968.

[Eid95]   E. Eid, A. Dickinson, D. Inglis, B. Ackland, E. R. Fossum, "A 256x256 CMOS Active Pixel Image Sensor," *Proc. of SPIE,* vol. 2415, February 1995.

[FairchildURL1]   "CCD486 Datasheet" http://www.fairchildimaging.com/

[Farrier97]   M. Farrier *et al.,* "Design and Processing Aspect of a 50 Megapixel Full Frame CCD I';mage Sensor," *IEEE Workshop on CCDs and Advanced Image Sensors,* Bruges, Belgium, 5-7 June 1997.

[Forza07]   04 June, 2007, "Forza Silicon Announces High Definition CMOS Imaging Technology", News release. (www.forzasilicon.com)

[Fossum93a]   E.R. Fossum, "Active pixel sensors challenge CCDs", *Laser Focus World,* pp. 83-87, June 1993.

[Fossum93b]   E. R. Fossum, "Active Pixel Sensors: Are CCD's Dinosaurs," *Proc. of SPIE,* vol. 1990, pp. 2-14, Feb. 1993.

[Fossum95]   E. R. Fossum, " CMOS image sensors: electronic camera on a chip," *IEDM Technical Digest, International Electron Devices Meeting,* pp. 17-25, 10-13 Dec. 1995

[Fossum98]   E. R. Fossum, *et al.,*"A 37×28 mm^2 600k-pixel CMOS APS dental X-ray camera-on-a-chip with self-triggered readout," *Digest of Technical Papers, IEEE International Solid-State Circuits Conference (ISSCC),* vol. 39, pp. 172-173, 1996.

[Fossum99]   E. R. Fossum, "Quantum Efficiency Improvements in Active Pixel Sensors," *US Patent No. 6,005,619,* 1999.

[FossumPC]   *Private communication with* E. R. Fossum, 2002.

[Foveon00]   URL: http://www.foveon.net, Foveon Inc., "Foveon Demonstrates 16 Megapixel CMOS Image Sensor", *PRESS RELIESE,* Cologne, Germany, Sept. 2000.

[Furumiya00]   M. Furumiya, H. Ohkubo, Y. Muramatsu, S. Kurosawa, Y. Nakashiba, "High sensitivity and no-cross-talk pixel technology for embedded CMOS image sensor," *IEDM Technical Digest, International Electron Devices Meeting,* pp. 701–704, 10-13 Dec. 2000.

[Ghazi00]   A. Ghazi, H. Zimmermann, P.Seegebrecht,"CMOS Photodiode With Enhanced Responsivity for the UV/Blue Spectral Range," IEEE Trans. on Electron Devices, vol. 49, no. 7, pp.1124-1128, July 2000.

[Groom99]   D. E. Groom, S. E. Holland, M. E. Levi, N. P. Palaio, S. Perlmutter, R. J. Stover, M. Wei, "Quantum Efficiency of a Back-illuminated CCD Image sensor: An Optical Approach," *Proc. of SPIE,* vol. 3649, pp. 80-90, 1999.

[Groom00]   D. E. Groom, *et al.,*"Recent progress on CCDs for astronomical imaging," *Proc. of SPIE,* vol. 4008, pp. 634-645, 2000.

[Guidash 97]   R. M. Guidash, T. H. Lee, P. P. K. Lee, D. H. Sackett, C. I. Drowley, M. S. Swenson, L. Arbaugh, R. Hollstein, F. Shapiro, S. Domer, "A 0.6 μm CMOS pinned photodiode color image sensor technology," *IEDM Technical Digest, International Electron Devices Meeting*, pp. 927-929, 7-10 Dec. 1997.

[Guidash00]   R. M. Guidash, *et al.*,"Active pixel image sensor with shared amplifier read-out," *US Patents 6,107,655*, August 2000.

[Hu03]   H. Hu, C. Zhu, Y. F. Lu, Y. H. Wu, T. Liew, M. F. Li, B. J. Cho, W. K. Choi, N. Yakovlev, "Physical and electrical characterization of HfO2 metal–insulator–metal capacitors for Si analog circuit applications," *Journal of Applied Physics*, vol. 94(1), pp. 551-557, July 1, 2003.

[Hung90]   K. K. Hung, P. K. Kuo, C. Hu, Y. C. Cheng, "A unified model for the flicker noise in metal-oxide-semiconductor field-effect transistors," *IEEE Transactions on Electron Devices*, vol. 37, pp. 654–665, March 1990.

[Hurwitz01]   J. Hurwitz, *et al.*, "A miniature imaging module for mobile applications," *Digest of Technical Papers, IEEE International Solid-State Circuits Conference (ISSCC)*, pp. 90-91, 5-7 Feb. 2001.

[Hynecek79]   J. Hynecek, "Virtual Phase CCD Technology," *IEDM Technical Digest, International Electron Devices Meeting,* Washington, DC, Dec. 3-5, 1979.

[Hynecek80]   J. Hynecek, "Virtual phase charge transfer device," *U.S. Pattent No. 4,229,752,* Oct. 1980.

[Hynecek81]   J. Hynecek, "Virtual phase technology: A new approach to fabrication of large area CCDs," *IEEE Transactions on Electron Devices*, vol. ED-28, no. 5, 1981.

[Hynecek92]   J. Hynecek, "CCM-A new low-noise charge carrier multiplier suitable for detection of charge in small pixel CCD image sensors," *IEEE Transactions on Electron Devices*, vol. 39, pp. 1972–1975, Aug. 1992.

[Hynecek01]   J. Hynecek, "Impactron—A new solid state image intensifier," *IEEE Transactions on Electron Devices*, vol. 48, pp. 2238–2241, Oct. 2001.

[Ihara98]   H. Ihara, *et al.*, "A 3.7x3.7um2 square pixel CMOS image sensor for digital still camera application," *Digest of Technical Papers, IEEE International Solid-State Circuits Conference (ISSCC)*, pp. 182-183, 5-7 Feb. 1998.

[ImagerLabsURL1]   "IL-C2004 Datasheet" http://www.imagerlabs.com/

[Inoue99]   I. Inoue, *et al.*, "New LV-BPD(Low Voltage Buried Photo-Diode) for CMOS Image sensor," *IEDM Technical Digest, International Electron Devices Meeting*, no.36.5, pp. 883-886, Dec. 1999.

[Inoue03]   I. Inoue, N. Tanaka, H. Yamashita, T. Yamaguchi, H. Ishiwata, H. Ihara, "Low-Leakage-Current and Low-Operating-Voltage Buried Photodiode for a CMOS Image sensor", *IEEE Transactions on Electron Devices*, vol. 50, no. 1, pp. 43-47, January 2003.

[Iversen03]   S. Iversen, *et al.* "An 8.3-Megapixel, 10-bit, 60 fps CMOS APS," *IEEE Workshop on CCDs and Advanced Image Sensors*, Elmau, Germany, 15-17 May 2003.

[Iwane07]   M. Iwane, et al., "52 Mega-pixel APS-H-size CMOS Image Sensor for Super High Resolution Image Capturing," *2007 International Image Sensor Workshop*, pp. 297-298, Ogunquit, Maine, June 7-10 2007.

[Janesick92]   J. Janesick, "Open-pinned phase charge couple device," *NASA Tech. Briefs*, vol. 16, no. 1, pp. 16, 1992.

[Janesick94a]   J. Janesick, "CCD with a 'thin gate'," *Laser Tech. Briefs*, vol. 2, no. 2, pp. 18, 1994.

[Janesick94b]   J. Janesick, "Frontside illuminated charge-couple device with high sensitivity to the blue ultraviolet and soft x-ray spectral range," *U.S. Pattent No. 5,365,092*, Nov. 1994.

[Janesick01]   J. Janesick, *Scientific Charge-Coupled Devices, SPIE*, Bellingham, WA, 2001.

[Jerram01]   P. Jerram, P. Pool, R. Bell, D. Burt, S. Bowring, S. Spencer, M. Hazelwood, I. Moody, N. Catlett, P. Heyes, "The LLLCCD: Low light imaging without the need for an intensifier," *Proc. of SPIE*, vol. 4306, pp. 178–186, 2001.

[Keenan85]   W. Keenan, D. Harrison, "A tin oxide transparent-gate buried channel virtual-phase CCD image sensor," *IEEE Transactions on Electron Devices*, vol. ED-32, no. 8, 1985.

[Kemeny97]   S. E. Kemeny, R. Panicacci, B. Pain, L. Matthies, E. R. Fossum, "Multiresolution image sensor," *IEEE Transactions Circuits and Systems for Video Technology*, vol. 7, no. 4, pp. 575-583, Aug. 1997.

[Kleinfelder01] S. Kleinfelder, L. SukHwan, L. Xinqiao, A. El Gamal, "A 10000 frames/s CMOS digital pixel sensor," *IEEE Journal of Solid-State Circuits*, vol. 36, no. 12, pp. 2049–2059, Dec. 2001.

[Koch 95]   C. H. Koch, H. Li (Eds), "Vision Chips: Implementing Vision Algorithms with Analog VLSI Circuits," *IEEE Computer Press*, 1995.

[KodakURL1]   "KAF-22000CE Datasheet" http://www.kodak.com/

[KodakURL3]   "KAF-4320E Datasheet" http://www.kodak.com/

[KodakURL2]   "KAF-16802CE Datasheet" http://www.kodak.com/

[Kreider95]   G. Kreider, J. Bosiers, B. Dillen, J. van der Heijden, W. Hoekstra, A. Kleimann, P. Opmeer, J. Oppers, H. Peek, R. Pellens, A. Theuwissen, "An mK x nK Modular Image Sensor Design," *IEDM Technical Digest, International Electron Devices Meeting*, pp.155-158, 1995.

[Kristianpoller64] N. Kristianpoller, D. Dutton, "Optical properties of 'liumogen' : A phosphor for wavelength conversion," *Applied Optics*, vol. 3, no. 2, 1964.

[Korthout03]   L. Korthout, *et al.* "A low power single chip VGA camera," *IEEE Workshop on CCDs and Advanced Image Sensors*, Elmau, Germany, 15-17 May 2003.

[Krymsky99]   A. Krymsky, D. VanBlerkom, A. Andersson, N. Bock, B. Mansoorian, E. R. Fossum, "A high speed, 500 frames/s, 1024×1024 CMOS active pixel sensor," *VLSI Circuits, 1999. Digest of Technical Papers 1999 Symposium*, pp. 137-138, 17-19 June 1999.

[Krymsky01]   A. Kyrimsky, N. Bock, D. VanBlerkom, N. Tu, E. R. Fossum ,"A high speed, 240 frames/s 4 megapixel CMOS sensor," *IEEE Workshop on CCDs and Advanced Image Sensors*, pp.28-31, Crystal Bay, Nevada, June 2001.

| [Lauxtermann99] | S. Lauxtermann, P. Schwider, P. Seitz, H. Bloss, J. Ernst, H. Firla, "A high speed CMOS image sensor acquiring 5000 frames/sec," *IEDM Technical Digest, International Electron Devices Meeting,* pp. 875-878, 5-8 Dec. 1999. |

[Lee95] P. Lee, R. Gee, M. Guidash, T. Lee, E. R. Fossum, "An active pixel sensor fabricated using CMOS/CCD process technology," *IEEE Workshop on CCDs and Advanced Image Sensors*, Dana Point, CA, 20-22 April 1995.

[Lee97] P. Lee, R. M. Guidash, *et al.*, "Active Pixel Sensor integrated with a Pinned Photodiode", *US Patent No. 5,625,210*, Apr. 29, 1997.

[LeeJ01] J. S. Lee and R.I Hornsey "CMOS Photodiodes with Substrate Openings for Higher Conversion Gain in Active Pixel Sensors" *IEEE Workshop on CCDs and Advanced Image Sensors*, pp. 173-175, Crystal Bay, Nevada, June 2001.

[LeeJ03] J. S. Lee, R. I. Hornsey, D. Renshaw, "Analysis of CMOS Photodiodes—Part II: Lateral Photoresponse," IEEE Trans. on Electron Devices, vol. 50, no. 5, pp. 1239-1245, May 2003.

[Lesser97] M. Lesser, D. Ouellete, A. J. P. Theuwissen, K. L. Kreider, H. Michaelis, "Packaging and operation of Philips 7kx9k CCDs," *IEEE Workshop on CCDs and Advanced Image Sensors*, Bruges, Belgium, 5-7 June 1997.

[Loinaz98] M. J. Loinaz, K. J. Singh, A. J. Blanksby, D. A. Inglis, K. Azadet, B. D. Ackland, "A 200-mW, 3.3-V, CMOS color camera IC producing 352×288 24-b video at 30 frames/s," *IEEE Journal of Solid-State Circuits*, vol. 33, no. 12, pp. 2029-2103, Dec. 1998.

[Loose01] M. Loose, *et al.*, "2/3" CMOS imaging sensor for high definition television," *IEEE Workshop on CCDs and Advanced Image Sensors*, pp.44-47, Crystal Bay, Nevada, June 2001.

[Madan83] S. K. Madan, B. Bhaumik, J. M. Vasi, "Experimental observation of avalanche multiplication in charge coupled devices," *IEEE Transactions on Electron Devices*, vol. ED-30, pp. 694–699, June 1983.

[Malinovich99] Y. Malinovich, E. Koltin, D. Cohen, M. Shkury, M. Ben-Simon, "Fabrication of CMOS Image Sensors," *Proc. of SPIE,* vol. 3649, pp. 212-218, 1999.

[Mann91] J. Mann, "Implementing early visual processing in analog VLSI: light adaptation in Visual information processing: from neurons to chips," *Proc. of SPIE,* vol. 1473, pp. 128-132, 1991.

[Mansoorian99] B. Mansoorian, Y. Horng-Yue, S. Huang, E. Fossum, "A 250 mW, 60 frames/s 1280×720 pixel 9 b CMOS digital image sensor," *Digest of Technical Papers, IEEE International Solid-State Circuits Conference (ISSCC)*, pp. 312-313, 15-17 Feb. 1999.

[Mead85] C. Mead, "A sensitive electronic photoreceptor," *1985 Chapel Hill Conference on VLSI*, pp. 463-471, 1985.

[Meisenzahl00] E. Meisenzahl, W. Chang, W. DesJardin, S. Kosman, J. Shepherd, E. Stevenson, K. Wong, "A 3.2 million pixel full-frame true 2-phase CCD image sensor incorporating transparent gate technology," *Proc. of SPIE,* vol. 3965, pp. 92-100, 2000.

[Melen73] R. Melen, "The tradeoff in monolithic image sensors: MOS vs. CCD," *Electronics*, vol. 46, pp. 106-111, May 1973.

236

[Mendis93a]   S. Mendis, S. Kemeny, E. R. Fossum, "A 128x128 CMOS active pixel image sensor for highly integrated imaging systems," *IEDM Technical Digest, International Electron Devices Meeting,* pp. 583-586, 1993.

[Mendis93b]   S. Mendis, B. Pain, R. Nixon, E. R. Fossum, "Design of a low-light-level image sensor with an on-chip sigma-delta analog-to-digital conversion," *Proc. of SPIE,* vol. 1900, pp. 31-39, 1993.

[Mendis94a]   S. Mendis, S. E. Kemeny and E. R. Fossum, "CMOS active pixel image sensor," *IEEE Transactions on Electron Devices,* vol. 41(3), pp. 452-453, 1994.

[Mendis94b]   S. Mendis, S. E. Kemeny, R. Gee, B. Pain, and E. R. Fossum, "Progress in CMOS active pixel image sensors," *Proc. of SPIE,* vol. 2172, pp. 19-29, 1994.

[Mendis97a]   S. Mendis, S. E. Kemeny, R. C. Gee, B. Pain, Q. Kim, and E. R. Fossum, "CMOS active pixel image sensors for highly integrated imaging systems," *IEEE Journal of Solid-State Circuits,* vol. 32(2), pp. 187-197, 1997.

[Mendis97b]   S. Mendis, A. Budrys, J. Lin, K. Cham, "Active Pixel Image Sensor in 0.35um CMOS Technology," *IEEE Workshop on CCDs and Advanced Image Sensors,* Bruges, Belgium, 5-7 June 1997.

[Meynants01]   G. Meynants, *et al.* "Fixed pattern noise suppression by differential readout chain for a radiation-tolerant image sensor," *IEEE Workshop on CCDs and Advanced Image Sensors,* pp. 56-59, Crystal Bay, Nevada, June 2001.

[Meynants03]   G. Meynants, *et al.* "A 35mm 13.89 million pixel CMOS active pixel image sensor *"IEEE Workshop on CCDs and Advanced Image Sensors,* Elmau, Germany, 15-17 May 2003.

[Muramatsu93]   M. Muramatsu, N. Suyama, K. Yamamoto, "UV Response of backside illuminated CCDs," *IEEE Workshop on CCDs and Advanced Image Sensors,* Waterloo, Canada, 9-11 June 1993.

[Nakamura 95]   J. Nakamura, S. E. Kemeny and E. R. Fossum, "A CMOS active pixel image sensor with simple floating gate pixels," *IEEE Transactions on Electron Devices,* vol. ED-42(9), pp. 1693-1694, 1995.

[Nixon96]   R. H Nixon, S. E. Kemeny, B. Pain, C. O. Staller and E. R. Fossum, "256x256 CMOS active pixel sensor camera-on-a-chip," *IEEE Journal of Solid-State Circuits,* vol. 31(12), pp. 2046-2050, 1996.

[Noble68]   P. Noble, "Self-scanned silicon image detector arrays," *IEEE Transactions on Electron Devices,* vol. ED-15(4), pp. 202-209, 1968.

[Oba97]   E. Oba, K. Mabuchi, Y. Lida, N. Nakamura, H. Miura, "A 1/4 inch 330k square pixel progressive scan CMOS active pixel image sensor," *Digest of Technical Papers, IEEE International Solid-State Circuits Conference (ISSCC),* pp. 180-181, 6-8 Feb. 1997.

[Olsen97]   B. H. Olsen, T. Shaw, B. Pain, R. A. Paniccaci, B. Mansoorian, R. H. Nixon, E. R. Fossum, "A Single Chip CMOS APS Digital Camera," *IEEE Workshop on CCDs and Advanced Image Sensors,* Bruges, Belgium, 5-7 June 1997.

[Pedrotti93]   F. L. Pedrotti, L. S. Pedrotti, *Introduction to Optics,* Prentice Hall, New Jersey, 1993.

[Photobit96]   Photobit LLC, *Prometheus Test Plan*, 7 March 1996.

[Ramacher99]   U. Ramacher, I. Koren, H. Geib, C. Heer, T. Kodytek, J. Werner, J. Dohndorf, J. U. Schlussler, J. Poidevin, S. Kirmser, "Single-chip video camera with multiple integrated functions," *Digest of Technical Papers, IEEE International Solid-State Circuits Conference (ISSCC)*, pp. 306-307, 15-17 Feb.1999.

[REFL1]   *Reference list:* [Wuu00] [Furumiya00] [Wuu01] [Chien02] [Takayanagi03] [Cho01] [Hurwitz01] [Cho00] [Mansoorian99] [Ramacher99] [Krymsky99] [Korthout03] [Meynants01] [Krymsky01] [Loose01] [Berezin01] [Meynants03]

[REFL2]   *Reference list:* [Mendis93a] [Mendis93b] [Wong97] [Weng96] [Blanksby97] [Olsen97] [Dickinson95] [Dickinson96] [Blanksby00] [Cheng03] [Loinaz98] [Mendis97] [Eid95] [Ackland96] [Yang01] [Kleinfelder01] [Kemeny97] [Zhou97]

[Ricquier92]   N. Ricquier, B. Dierickx, "Pixel structure with logarithmic response for intelligent and flexible image sensor architectures," *Microelectronic Engineering*, vol. 19, pp. 631-634, 1992.

[Ricquier95]   N. Ricquier, B. Dierickx, "Active pixel CMOS image sensor with on-chip non-uniformity correction," *IEEE Workshop on CCDs and Advanced Image Sensors*, Dana Point, CA, 20-22 April 1995.

[Scheffer01]   D. Scheffer, *et al.* "A 6.6Mpixel CMOS image sensor for electrostatic PCB inspection," *IEEE Workshop on CCDs and Advanced Image Sensors*, pp. 145-148, Crystal Bay, Nevada, June 2001.

[Shortes74]   S. Shortes, W. W. Chan, W. C. Rhines, J. B. Barton, D. R. Collins, "Characteristics of thinned backside illuminated charge couple device image sensors," *Applied Physics Letters*, vol. 24, no. 11, pp. 565-567, 1974

[Sheu87]   B. J. Sheu, D. L. Scharfetter, P. K. Ko, and M. C. Jeng, "BSIM: Berkeley short-channel IGFET model for MOS transistors," *IEEE Journal of Solid-State Circuits*, vol. SSC-22, pp. 558–565, 1987.

[Takayanagi03]   I. Takayanagi, M. Shirakawa, K. Mitani, M. Sugawara, S. Iversen, J. Moholt, J. Nakamura, R. R. Fossum, "A 1 1/4 inch 8.3M pixel digital output CMOS APS for UDTV application," *Digest of Technical Papers, IEEE International Solid-State Circuits Conference (ISSCC)*, vol.1, pp. 216-217, 9-13 Feb. 2003.

[Teranishi82]   N. Teranishi, *et al.*, "No image lag photodiode structure in the in-line CCD image sensor," *IEEE Electron Devices Magazine*, pp. 324-327, 1982.

[Thornber74]   K. K. Thornber, "Theory of noise in charge-transfer devices," *Bell Syst. Tech. J.*, vol. 53, no. 7, pp. 1211-1262, 1974.

[Tian00]   H. Tian, A. El Gamal, "Analysis of 1/f Noise in CMOS APS," *Proc. of SPIE*, vol. 3965, pp. 168-176, 2000.

[Tower99]   Tower Semiconductor Inc., News Archive, August 2, 1999, www.towersemi.com

[Theuwissen95]   A. J. P. Theuwissen, *Solid-State Imaging with Charge-Coupled Devices*, Kluwer Academic Publishers, May 1995.

[Turley92]   A. Turley, B. Frias, A. Santos, R. Tacka, L. Colquitt, M. Polinsky, J. Ebner, E. Juergensen, R. Bishop, P. Difonzo, "Design and fabrication of tin oxide gate CCD visible imaging arrays," *Proc. of SPIE,* vol. 1693, pp.113-121, 1992.

[URL1]   BASF Corporation URL: http://www.basf.com, web page.

[URL2]   Roper Scientific Corporation URL : http://www.photomet.com,

[URL3]   Texas Instruments: TC253SPD-30 CCD datasheet, [On-line] http://focus.ti.com/docs/prod/folders/print/tc253spd-30.html

[URL4]   (Marconi Applied) e2v Technologies: datasheet CCD65 [Online] http://e2vtechnologies.com/product_guides/l3vision_sensors.htm

[Yadid-Pecht91]   O. Yadid-Pecht, R. Ginosar, Y. Shacham-Diamand, "A random access photodiode array for intelligent image capture," *IEEE Transactions on Electron Devices*, vol. 38(8), pp. 1772-1781, 1991.

[Yang98]   G.Yang, O. Yadid-Pecht, C. Wrigley, B. Pain, "A snap-shot CMOS active pixel image sensor for low-noise, high-speed imaging," *IEDM Technical Digest, International Electron Devices Meeting,* pp. 45–48, 6-9 Dec. 1998.

[Yang01]   G. Yang, C. Sun, C. Wrigley, D. Stack, C. Kramer, B. Pain, "Dynamically reconfigurable imager for real-time staring vision systems," *IEEE Workshop on CCDs and Advanced Image Sensors*, pp. 20-23, Crystal Bay, Nevada, June 2001.

[Velghe93]   R. M. D. A. Velghe, D. B. M. Klaassen, F. M. Klaassen, "Compact modeling for analogue circuit simulation," *IEDM Technical Digest, International Electron Devices Meeting,* pp. 485-488, 5-8 Dec. 1993.

[Weckler67]   G. P. Weckler, "Operation of p-n junction photodetectors in a photon flux integration mode," *IEEE Journal of Solid-State Circuits*, vol. SC-2, pp. 65-73, 1967.

[Wen99]   D. Wen, R. Bredthauler, P. Bates, P. Vy, R. Potter, "Performance characteristics of a 9216x9216 pixel CCD," *IEEE Workshop on CCDs and Advanced Image Sensors*, Karuizawa, Nagano, Japan, 10-12 June 1999.

[Weng96]   H.-S. Weng, R.T. Chang, E. Crabbe, P. Agnello, "CMOS active pixel image sensors fabricated using a 1.8 V, 0.25 µm CMOS technology," *IEDM Technical Digest, International Electron Devices Meeting,* pp. 915–918, 8-11 Dec. 1996.

[White74]   M. H. White, D. R. Lampe, F. C. Blaha, I. A. Mack, "Characterization of surface channel CCD image arrays at low light levels," *IEEE Journal of Solid-State Circuits*, vol. SC-9, pp. 1-14, 1974.

[Wong97]   H.-S. P. Wong, "CMOS image sensors-recent advances and device scaling considerations," *IEDM Technical Digest, International Electron Devices Meeting,* pp. 201-204, 7-10 Dec. 1997.

[Wuu00]   S.-G. Wuu, D.-N. Yaung, C.-H. Tseng, H.-C. Chien, C. S. Wang, Y.-K. Hsiao, C.-K. Chang, B. J. Chang, "High performance 0.25-um CMOS color imager technology with non-silicide source/drain pixel," *IEDM Technical Digest, International Electron Devices Meeting,* pp. 705-708, 10-13 Dec. 2000.

[Wuu01]   S.-G. Wuu, H.-C. Chien, D.-N. Yaung, C.-H. Tseng, C. S. Wang, C.-K. Chang, Y.-K. Hsaio, "A high performance active pixel sensor with 0.18um CMOS color imager technology," *IEDM Technical Digest, International Electron Devices Meeting,* pp. 24.3.1-24.3.4, 2-5 Dec. 2001.

[Zhou97]   Z. Zhimin, B. Pain, E. R. Fossum, "Frame-transfer CMOS active pixel sensor with pixel binning," *IEEE Transactions on Electron Devices*, vol. 44, no. 10, pp. 1764, Oct. 1997.

INDEX